Climate Change Risks and Adaptation

LINKING POLICY AND ECONOMICS

Please cite this publication as:
OECD (2015), *Climate Change Risks and Adaptation: Linking Policy and Economic*, OECD Publishing, Paris.
http://dx.doi.org/10.1787/9789264234611-en

ISBN 978-92-64-23460-4 (print)
ISBN 978-92-64-23461-1 (PDF)

Photo credits: © ALCE-Fotolia.com.

Corrigenda to OECD publications may be found on line at: *www.oecd.org/publishing/corrigenda*.

Foreword

Climate Change Risks and Adaptation: Linking Policy and Economics *provides practical guidance and the latest evidence for policy makers on how they can more reliably estimate the costs and benefits of adaptation. The report is intended as a guide for policy makers to better identify, characterise and address climate risks, helping them mainstream adaptation into decision making by taking into account a range of uncertainties. These measures should help inform policies, prioritise scarce resources and unlock sufficient finance for adaptation.*

Climate change is a daunting challenge that poses major risks to our economies, societies and environment. Increasing average temperatures could be accompanied by more frequent and intense extreme weather and rising sea levels, among other developments. Recent events illustrate how countries could be affected by climate change. The summer of 2003 was the hottest experienced in Europe since the 1500s. In France, it led to 15 000 excess heat-related deaths. In 2012, sea level rise exacerbated the flooding of the east coast of the United States resulting from Hurricane Sandy, contributing to the estimated total damages of USD 65 billion. As global average temperatures rise, so will the magnitude of climate-related risks.

The most severe risks can only be avoided if the world moves, in the second half of the century, towards the complete elimination of greenhouse gas emissions from the combustion of fossil fuels. However, we also need to address the risks resulting from our past actions. Past emissions, and the future emissions locked-in by legacy infrastructure, mean that the climate will continue to change even if future investment decisions start to move away from fossil fuels.

As we move into an increasingly uncertain future, it is vital that countries strengthen their ability to understand, plan for and continuously manage these climate risks. The trends are encouraging: more and more OECD countries are developing national strategies to cope with climate change, and sub-national and private actors are also increasingly tackling this issue. However, there is a pressing need to translate planning into implementation and to deal with the adverse effects that our past and current actions will have on our planet's nature.

This report takes stock of adaptation efforts in OECD countries to date, and points to where renewed efforts will be required. It underscores the message that successfully preparing for the changes brought by future climate change will also entail looking beyond monetary valuation and ensuring that the expectations and needs of affected communities are taken into account.

The OECD will continue to assist member and partner countries as the challenges derived from climate change and the need to take decisive actions become more evident.

Angel Gurría

OECD Secretary General

Acknowledgements

Climate Change Risks and Adaptation: Linking Policy and Economics is an output of the OECD's Environment Directorate.

This report (except for Chapters 3 and 6) was written by Michael Mullan, Takayoshi Kato, Nicolina Lamhauge and Jennifer Helgeson, under the supervision of Simon Buckle. It benefitted from substantive contributions by Britta Labuhn, Mikaela Rambali, Darina Petrova and Lola Vallejo. Chapter 3 was written by Paul Watkiss, Alistair Hunt, Josselin Rouillard, Jenny Troeltzsch and Manuel Lago. Contributors to that chapter are: Jeroen Aerts, Anne Biewald, Francesco Bosello, Aline Chiabai, Areti Kontogianni, Ekko van Ierland, Ibon Galarraga, Petr Havlik, Onno Kuik, Reinhard Mechler, Elisa Sainz de Murieta, Paolo Scussolini and Michalis Skourtos. Chapter 6 was written by Alistair Hunt and Paul Watkiss and benefitted from contributions by Federica Cimato.

The authors would like to thank Simon Buckle, Gerard Bonnis, Nils-Axel Braathen, Jan Corfee-Morlot, Kate Eklin, Jane Ellis, Catherine Gamper, Ada Ignaciuk, Elisa Lanzi, Xavier Leflaive and Robert Youngman at the OECD for their helpful input and feedback. The authors are also grateful for the oversight, review and comments by the Working Party on Climate, Investment and Development (WPCID) and the Environment Policy Committee (EPOC). The production benefitted from the assistance of Sama Al-Taher Cucci, Ruth Catts-Lemaire, Katerina Rus, Gabriella Scaduto-Mendola, Stéphanie Simonin-Edwards and Delphine Versini. Katherine Kraig-Ernandes and Janine Treves provided valuable editorial support.

Financial support from the US Environmental Protection Agency (EPA) is gratefully acknowledged. Chapters 3 and 6 draw on the research, analysis and review of the ECONADAPT project, funded by the European Union's Seventh Framework Programme for research, technological development and demonstration and the co-funding provided for this project by the UK Department for International Development and the International Development Research Centre. Additional information, including more detailed estimates on the costs and benefits of adaptation (Chapter 3), is provided in a supporting ECONADAPT report (2015) on the Costs and Benefits of Adaptation (*http://econadapt.eu/*). The views expressed in Chapters 3 and 6 do not necessarily reflect those of the ECONADAPT funders.

Table of contents

Acronyms ... 8
Executive summary ... 9

Chapter 1. Risks in a changing climate ... 13
Key messages ... 14
Adaptation through risk management ... 14
The nature of a changing climate ... 15
Mounting costs of climate impacts ... 16
Climate risks and uncertainties ... 18
Notes ... 19
References ... 19

Chapter 2. Approaches to managing climate risks ... 21
Key messages ... 22
National measures to address climate change risks ... 22
The role of sub-national governments in addressing climate risks ... 27
Private sector engagement in adaptation ... 29
Addressing the social aspect of adaptation ... 31
The importance of measuring progress in addressing climate risks ... 32
References ... 33

Chapter 3. Overview of costs and benefits of adaptation at the national and regional scale ... 37
Key messages ... 38
The costs and benefits of adaptation ... 38
From theory to practice ... 39
Current state of the literature ... 42
Global estimates ... 44
National estimates ... 45
Risk- and sector-based estimates ... 47
Discussion of the current state of evidence and key gaps ... 63
Notes ... 64
References ... 64

Chapter 4. Framing risk-based approaches to adaptation planning ... 77
Key messages ... 78
A framework for a risk-based approach to adaptation planning ... 78
i) Identifying risks ... 79
ii) Characterising risks to be addressed ... 83
iii) Choosing and exploring policy options ... 89

iv) Feedback process . . . 91
References . . . 92

Chapter 5. **Financing adaptation in OECD countries** . . . 97
Key messages . . . 98
Framework for understanding the budgetary impacts of climate change adaptation . . . 98
Financing adaptation at the sub-national level . . . 107
Financing adaptation at the sectoral level . . . 109
References . . . 112

Chapter 6. **Tools to mainstream adaptation into decision-making processes** . . . 117
Key messages . . . 118
Adaptation mainstreaming . . . 118
The context for adaptation mainstreaming . . . 119
Mainstreaming entry points and examples . . . 122
Enabling conditions for mainstreamed decision making . . . 124
Decision support tools for mainstreaming . . . 128
Appraisal review findings and lessons . . . 131
Notes . . . 135
References . . . 135

Tables

2.1. Coverage of adaptation in National Communications . . . 23
2.2. Examples of local level approaches to adaptation planning . . . 28
2.3. Factors influencing climate change adaptation in the private sector . . . 31
3.1. Updated quality of the coverage of the sectors in the adaptation literature . . . 42
4.1. Key questions for identifying risks . . . 80
4.2. Status of the evidence base for climate-related risks at the national level . . . 81
4.3. Types of adaptation measures to reduce risks . . . 90
5.1. Tools for government action to support private adaptation . . . 102
5.2. Climate change projections of insured losses and/or insurance prices . . . 103
5.3. Main sources of funding and financial instruments for urban adaptation . . . 108
5.4. Examples of financing for adaptation at the sectoral level . . . 110
6.1. Examples of appraisal methods used in the adaptation context . . . 130
6.2. Attributes and application of decision support methods for adaptation . . . 132

Figures

1.1. Economic losses from climatological, meteorological and hydrological disasters, 1980-2014 . . . 17
1.2. Distribution of economic loss events worldwide 2014 . . . 17
3.1. National level adaptation cost studies . . . 43
4.1. Four steps in a risk-based approach . . . 79
4.2. Sectoral priorities for enhancing the evidence base . . . 81
4.3. Characterising risks . . . 83
4.4. Dynamics of risks over time . . . 85
4.5. Conceptual framework of a feedback process for a risk-based approach . . . 92

6.1. Iterative climate risks and adaptation 121
6.2. Mainstreaming steps and entry points..... 122
6.3. Decision support tools for adaptation 130
6.4. Applications of new decision support tools for adaptation..... 131

Acronyms

CAP	Common Agricultural Policy
CBA	Cost benefit analysis
CDP	Carbon Disclosure Project
CEA	Cost effectiveness analysis
COP	Conference of the Parties
DIVA	Dynamic Interactive Vulnerability Assessment
EACC	Economics of Adaptation to Climate Change
EEA	European Environment Agency
EIA	Environmental impact assessment
EU	European Union
I-A	Impact assessment
IAM	Integrated assessment model
IPCC	Intergovernmental Panel on Climate Change
IRM	Iterative risk management
MCA	Multi criteria analysis
OECD	Organisation for Economic Co-operation and Development
PA	Portfolio Analysis
PPP	Public-Private Partnerships
PSI	Private Sector Initiative
RDM	Robust decision making
ROA	Real options analysis
UNDP	United Nations Development Programme
UNFCCC	United Nations Framework Convention on Climate Change

Climate Change Risks and Adaptation
Linking Policy and Economics

Executive summary

The effects of climate change are already starting to be felt: sea-levels are rising, average temperatures are increasing and precipitation patterns are changing. Societies and ecosystems have always adapted to changes in their climate, but projected changes over the course of this century would lie outside of the bounds of historical experience. At the same time, the pace of the projected change is faster than any climate shift in the past millions of years. The consequences of this change will include shifts in agricultural production, changes in morbidity and losses of valuable ecosystems. Recent OECD modelling found that climate change impacts could reduce global GDP by between 1% and 3.2% per year by 2060. These are the known consequences. However, as the world moves further outside of its usual operating parameters, there is an increasing risk of encountering "severe, pervasive and irreversible impacts" according to the Intergovernmental Panel on Climate Change (IPCC).

In preparing for climate change, countries are faced with the challenge of responding to a broad range of uncertain risks. Climate projections are subject to numerous uncertainties, including the path of future emissions and the sensitivity of climate to concentrations of atmospheric greenhouse gases. Downscaling global changes to levels relevant for decision making brings in further challenges, due to data gaps and inherent challenges in modelling. Moreover, a further layer of uncertainty is added by the interaction between socio-economic trends and climate change. Relevant trends include: ageing populations, economic and social inequality and global trade patterns.

Information on the costs and benefits of adaptation is important for justifying and prioritising action, yet existing estimates only provide a partial picture. The evidence base on costs and benefits has significantly improved in recent years, as their sectoral and national coverage has improved. Increasingly, studies are focussing upon the needs of policy makers and incorporate more and more new elements such as transaction costs and the consideration of uncertainty. However, major gaps remain in sectoral coverage, as impacts on businesses and ecosystems remain poorly understood. Studies have predominantly focussed on monetary impacts, while coverage of non-market impacts (such as the social impacts) remains sparse. Few studies have examined the systemic impacts of climate change, as impacts in one area cascade through the economy. Lastly, the potential economic impacts of future climate extremes are not fully incorporated in existing estimates of the costs of climate change.

In addition to these scientific and economic factors, climate change also has strong normative and equity dimensions. Technical analysis provides an important input into this process, but it is also necessary to account for people's perceptions of the risks from climate change. In part, this is because decisions on how to respond to the risks from climate change frequently involve trade-offs between different values: for example, people's attachment to living in certain place versus the risk reduction that would be achieved if they were to relocate. In addition, the impacts of climate change will be felt most severely by the poor and the marginalised. Accounting for these disparate impacts can improve the quality and acceptability of the resulting decisions.

Recognising these challenges, and building on previous OECD analysis and countries' experiences, this publication proposes an iterative process for managing the risks from climate change:

- **Identifying risks:** Undertaking research and consultation to identify the range of potentially relevant risks that could arise from climate change.
- **Characterising risks:** Risks can be monitored, reduced, transferred or absorbed depending upon their characteristics. The appropriate responses will depend upon factors including the likely severity of consequences, the likelihood of the risk occurring, socio-economic consideration to address the risks, and the projected evolution of the risk. New decision-making approaches, such as real options analysis, can help to inform policy choices, as well as the design and financing of infrastructure.
- **Choosing and exploring policy responses:** New policies, or reforms to existing interventions, may be required to ensure that risks are reduced or transferred as appropriate. This can include early action to integrate adaptation into current decisions, or activities with long lifetimes, such as infrastructure or land use planning.
- **Feedback and learning:** On-going monitoring and regular evaluation can help to ensure that risk management measures are on track, implement adjustments if required and identify newly-emerging risks.

In moving from planning to implementation, there is a need to channel and mobilise sufficient public and private investment towards risk reduction activities. Estimates of the scale of resources required for risk reduction suggest that the availability of funding is not an absolute constraint in OECD countries. However, the challenge is to ensure that funds are available when and where they are needed. The challenges and constraints vary by sector but, in general, achieving increased investment in risk reduction will require enhanced collaboration between the public and private sectors. The public sector has an important role to play in facilitating this happens in an efficient and equitable manner.

As climate and socio-economic trends drive losses upwards, the treatment of risks that remain after risk transfer and risk reduction activities have taken place will become increasingly important. The public sector directly bears the costs of weather impacts through damage to its assets, funding to cover indemnities (whether statutory or implicit) and the indirect consequences of economic disruption. The scale of these potential public sector liabilities are not routinely assessed or managed by OECD countries. Improved evidence on these potential climate liabilities would clarify the costs of inaction on climate change, inform investments in risk reduction and support the financial management of adverse shocks.

In practice, national approaches to adaptation have primarily focused on integrating adaptation in existing plans and processes rather than introducing stand-alone policies or programmes. Such integration, however, requires the identification of suitable entry points in the policy process that reflect the specific sector and national contexts. The success of an integrated approach to adaptation will largely depend on the timing of such efforts and their ability to take advantage of existing opportunities and intervention points.

Recognising the possible limitations of existing appraisal processes due to the prevailing climate change uncertainty, new decision support tools have emerged. While some of these tools are resource-intensive and complex to use, "light-touch" approaches can capture the essence of these tools while balancing pragmatism with economic rigour.

Climate Change Risks and Adaptation
Linking Policy and Economics

Chapter 1

Risks in a changing climate

This chapter provides an overview of the risks that are expected to arise from climate change. It includes evidence of the rising costs of extreme weather events, both globally and within OECD countries. This chapter explores the role of uncertainty about future climate change and the implications for decision making.

Key messages

- Climate change is creating new risks and exacerbating existing ones. Ecosystems will shift, food production placed under increasing pressure and some types of extreme weather events will increase in frequency and severity. As global temperatures rise, the likelihood of encountering potentially catastrophic changes increases.
- Flexible, iterative risk management approaches can support decision making given uncertainty about the future. This is vital since the characteristics of climate risks are determined by complex, and often unpredictable, interactions between economic, social and environmental systems.
- Progress needs to be made on three fronts to address the risks from climate change: reducing greenhouse gas emissions; supporting risk reduction activities; and enhancing mechanisms to transfer and share risks.

Adaptation through risk management

Risks from climate variability and change cannot be completely eliminated, but the exposure and vulnerability of economies and societies to those risks can be reduced. For example, farmers may choose to adjust their farming practices in response to observed changes in the climate, while companies may adjust their production processes in the face of supply change disruptions. Investment in research and development can support the transition to a changing climate. Well-functioning insurance mechanisms, responsive public services and clear information can all help to reduce the negative consequences of the risks that materialise (OECD, 2014).

Households' and businesses' self-interest provides a strong incentive to manage risks and exploit any opportunities arising from climate change. However, adaptation actions by the private sector will be informed by the institutional and policy frameworks in place. For example, if industrial users are provided with water at subsidised rates, they will have less economic incentive to adjust water intensive practices. Governments therefore have an important role to play in creating an enabling environment for adaptation.

There are two, inter-related areas of risk management where the government will play an important role (Mechler et al., 2014):

- **Reducing climate risk:** Examples include addressing market failures, ensuring that the public sector's actions contribute to climate resilience and raising awareness of climate risks. Existing policies, such as those relating to water management and land-use planning, may need to be examined to check that they do not act as barriers to risk reduction.
- **Transferring climate risk:** Adaptation strategies can reduce, but often not eliminate the risks generated by a changing climate. Well-designed risk transfer and risk sharing arrangements can strengthen households' and businesses' ability to manage the economic impacts that occur, while retaining incentives for risk reduction.

The residual impacts that remain after risk reduction and risk transfer will be absorbed by households, businesses and the public sector. Policies to absorb the financial impacts of residual risks should be designed and implemented with particular consideration for the long-term efficiency of adaptation measures and equity.

The nature of a changing climate

The consequences of growing concentrations of greenhouse gases are becoming increasingly apparent. Temperatures are increasing, precipitation patterns are changing, snow and ice cover decreasing, and sea levels rising. The impact of these changes is becoming evident, primarily through increases in the intensity and frequency of extreme weather events such as heat waves, floods and wildfires. Trend changes, or "slow onset" events, that gradually evolve as a result of incremental changes in the climate will also have profound economic and social impacts. These include changes in the timing of seasonal life-cycle events for animal and plant species, agricultural shifts affecting food production processes, and the impact of precipitation and temperature changes on clean water availability and human well-being (IPCC, 2014a).

A particular concern with climate change is that rising temperatures increase the likelihood of encountering "pervasive, severe, and irreversible" impacts to natural and human systems (IPCC, 2014a). These impacts include the possibility of the Greenland ice sheet collapsing, the melting of the Himalayan icecap glaciers, and the die back of the Amazon rainforest (Dow et al., 2013). Although these are very unlikely to occur in the near-term, the consequences would be catastrophic in the long-term. For example, the Greenland ice sheet contains enough water to increase global sea levels by 7 metres (The ice2sea Consortium, 2013), while the Amazon rainforest is one of the world's major carbon dioxide sinks.

Substantial and sustained reductions in greenhouse gas emissions are required to limit the extent of climate change and reduce the likelihood of encountering severe, irreversible changes (IPCC, 2013). This needs to be accompanied with action to prepare for the effects of a changing climate, as the world is already committed to several decades of warming. This commitment is due to the greenhouse gas emissions that have already accumulated in the atmosphere and the path dependence of future emissions resulting from existing infrastructure and energy systems.

Box 1.1. **Definitions for adaptation, risk and resilience**

This report uses the following definitions of key terms, which are based on those proposed by the Intergovernmental Panel on Climate Change (IPCC):

- **Adaptation:** Adjustment in natural or human systems in response to actual or expected climatic stimuli or their effects, which moderates harm or exploits beneficial opportunities.
- **Exposure:** The presence of people, livelihoods, species or ecosystems, environmental functions, services, and resources, infrastructure, or economic, social, or cultural assets in places and settings that could be adversely affected.
- **Hazard:** The potential occurrence of natural or human-induced events that may have adverse impacts on human and natural systems.
- **Residual risk:** Risks that ongoing risk reduction processes have not mitigated, reduced sufficiently or eliminated.

Box 1.1. **Definitions for adaptation, risk and resilience** *(cont.)*

- **Risk:** Potential for consequences where something of value is at stake and where the outcome is uncertain, resulting from the interaction of climate-related hazards (gradual changes as well as extreme events) with the vulnerability and exposure of human and natural systems.
- **Vulnerability:** The propensity or predisposition to be adversely affected. Vulnerability encompasses a variety of concepts and elements including sensitivity or susceptibility to harm and lack of capacity to cope and adapt.

Source: IPCC (2014a), *Climate Change 2014: Impacts, Adaptation, and Vulnerability*, Contribution of Working Group II to the Fifth Assessment Report of the Intergovernmental Panel on Climate Change, Cambridge University Press, Cambridge and New York; IPCC (2012), *Managing the Risks of Extreme Events and Disasters to Advance Climate Change Adaptation*, A Special Report of Working Groups I and II of the Intergovernmental Panel on Climate Change, Cambridge University Press, Cambridge and New York.

Mounting costs of climate impacts

The economic and social costs associated with both slow and rapid onset weather events continue to set new records. This is in large part due to increasing exposure of people and economic assets to weather- and climate-related risks, a trend that is projected to continue both in OECD countries and globally (e.g. IPCC, 2012; Kuczinski and Irvin, 2012). Climate change is projected to become an increasingly important driver of losses in the future. Simulations by the Organisation for Economic Co-operation and Development (OECD) finds that if average global temperatures increase between 1.5 °C and 4.5 °C this would reduce global GDP by around 1.0% and 3.2% per year by 2060[1] (OECD, 2015 forthcoming). These average figures mask large regional and sectoral variations.

Figure 1.1 shows the trend of economic losses from climatological (e.g. heat waves, droughts and wildfires), meteorological (e.g. storms) and hydrological (e.g. floods) events between 1980 and 2014, based on data collected, and classifications used, by EM-DAT (n.d.).[2] These loss estimates only represent a subset of the full costs of weather events, due to data gaps and the inherent difficulty of capturing some types of impact (IPCC, 2012). Non-market impacts, such as the loss of ecosystem services, damage to cultural heritage and long-term health effects, are not routinely included in estimates of economic losses. Loss estimates do, nonetheless, provide an indicative trend.

Meteorological events are the largest source of recorded economic damages from natural disasters. In 2014, 46% of global economic disaster losses were caused by storms, 27% by hydrological events and 20% by climatological events according to data collected by Munich Re (2015). One of the most devastating windstorms to make landfall in an OECD country in recent years was Hurricane Katrina, which hit the United States in August 2005. Over 1 800 people lost their lives and economic losses accounted for over USD 100 billion, of which around USD 45 billion were insured on the private market (IPCC, 2007; Munich Re, 2006). Windstorm Christian, which hit Northern Europe in October 2013, caused insurance losses of an estimated USD 1.4 billion with a further USD 2 billion of economic losses (AON Benfield, 2014). The majority of the losses were concentrated in Denmark and Germany, but Netherlands, France, Latvia, Luxembourg, the United Kingdom and Sweden were also affected. Similarly, windstorm Xynthia, which crossed Europe in February and March 2010, caused insurance losses of around USD 3.65 billion, and economic losses amounting to around USD 4.5 billion (Liberato et al., 2013).

Figure 1.1. **Economic losses from climatological, meteorological and hydrological disasters, 1980-2014**

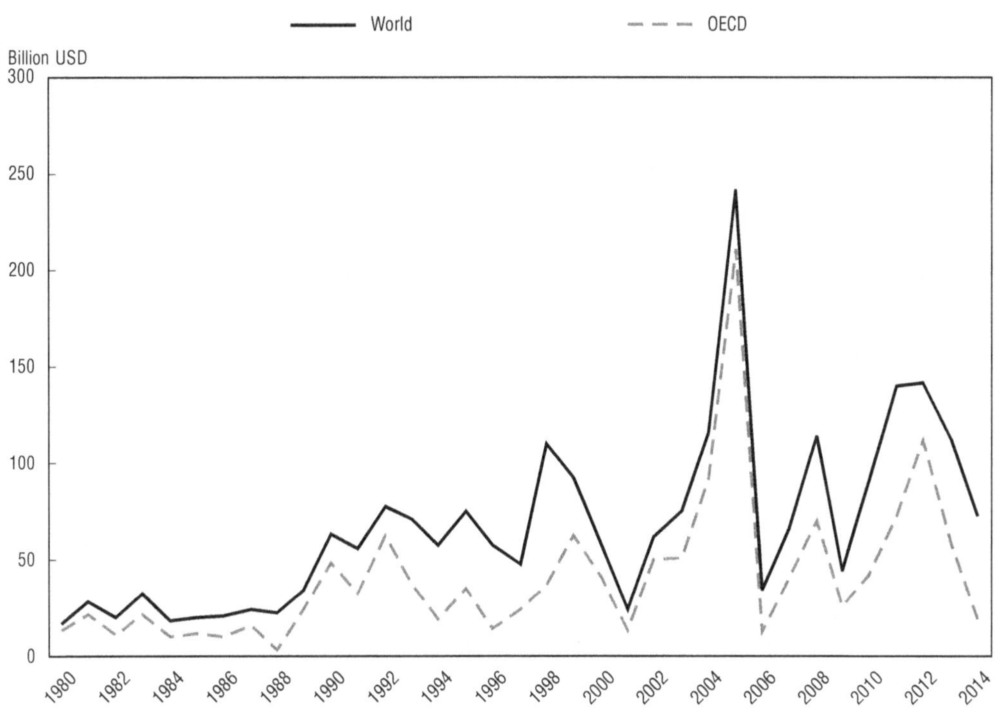

Source: EM-DAT (*Emergency Event Database*) (n.d.), "The International Disaster Database", Centre for Research on the Epidemiology of Disasters, *www.emdat.be/* (accessed 27 February 2015).

Figure 1.2. **Distribution of economic loss events worldwide 2014**

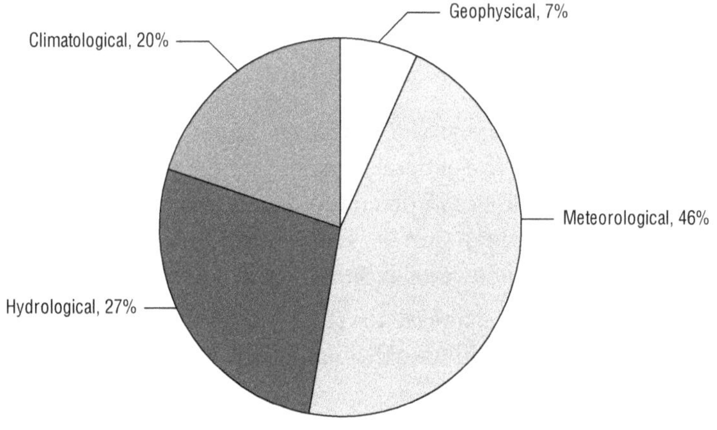

Source: Munich Re (2015), "Loss events worldwide 2014: Percentage distribution", *Annual statistics 2014*, *www.munichre.com/en/reinsurance/business/non-life/natcatservice/annual-statistics/index.html.*

Flooding is the second largest source of climate-related losses in OECD countries. In the EU, for example, the cost of flooding between 2000 and 2012 was on average around EUR 4.9 billion per year (Jongman et al., 2014). By 2050, this figure could be as high as EUR 23.5 billion per year, with two-thirds of this increase driven by economic growth and one-third by climate change (Jongman et al., 2014). In the Netherlands alone, the cost of implementing a comprehensive set of flood protection and flood risk management measures is projected to cost around EUR 1.2-1.6 billion per year up to 2050 and EUR 0.9-1.5 billion per year during the 2050-2100 period (ClimateCost, 2011). In Sweden, the investment needed across a range of sectors (e.g. transport, water treatment and infrastructure) have been estimated at EUR 10 billion in total over the period 2010-2100 (Swedish Commission on Climate and Vulnerability, 2007).

Slow onset events, such as the drought that has affected California since 2012, can also have substantial economic impacts. The combination of low precipitation and high temperatures has contributed to the most severe drought conditions in more than a century (Griffin and Anchukaitis, 2014). This drought has not been directly attributed to climate change, but provides an indication of the type of event that may occur more frequently due to climate change. In 2014 alone, the economic impacts of the drought on crop production, livestock and dairies, in addition to the costs of increased groundwater pumping was estimated at around USD 2.2 billion. On top of that, the drought is estimated to have contributed to the loss of 17 100 seasonal and part-time agricultural sector jobs in 2014 (Howitt et al., 2014).

Australia is also highly exposed to coastal flooding and sea level rise, with more than 75% of the population living on or near the coast. DCCEE (2011) estimated that commercial, industrial, road, rail, and residential assets worth more than AUD 226 billion (2008 replacement value) would be at risk of coastal flooding and erosion if sea levels were to rise by 1.1 metres by 2100 (high end scenario). Water supplies are another key area of concern, as Australia is the driest inhabited continent in the world. Although specific extreme weather events cannot be attributed to climate change, the frequency of heat waves experienced in parts of Australia has already surpassed levels previously predicted for 2030. The return period for annual extremes of maximum daily temperatures is projected to increase from once every 20 years to once every 2-5 years by the middle of the century. In the absence of significant cuts in greenhouse gas emissions, today's unusually hot weather may become the norm during the summer months (IPCC, 2012).

Climate risks and uncertainties

Projections of climate change are all uncertain, but some aspects of the climate are better understood than others. The general trend of rising temperatures is clear, but other climate alterations, such as changes in precipitation, are more difficult to project over the course of the century (IPCC, 2014a). Current knowledge of climate change can be divided into three categories (van Bree and van der Sluijs, 2014):

- **Statistical uncertainties** can be expressed in statistical or probabilistic terms. For example, climate models can be used to estimate the likelihood of a flood occurring in a given year (e.g. once in 100 years).
- **Scenario uncertainties** are those where the potential consequences can be understood, but the probabilities associated with them are not known. For example, the IPCC does not assign likelihoods to the different emissions scenarios, since the processes driving them are the result of peoples' choices and cannot be readily modelled.
- **"Ignorance" uncertainties** are those for which it is impossible to estimate the magnitudes in probabilistic or scenario-based terms, or to even assess the range of potential consequences.

A key challenge for adaptation planning is to prepare for potential impacts that are subject to these types of uncertainty. Scientific advances, improved data collection and better modelling methodologies can all help to reduce sources of uncertainty. However, the inherent unpredictability, randomness and chaotic behaviour of climate, human and nature systems mean that some uncertainties will always remain (van Bree and van der Sluijs, 2014). Ignoring these uncertainties can lead to inefficient responses to climate risks, such as over-investment in protective infrastructure. However, the perception that issues are too uncertain can paralyse decision making and incur costs of inaction, such as retrofitting or the need to relocate assets.

Focussing on climate risks enables scientific information and value judgements to be combined to assist decision making for an uncertain future. This approach targets the uncertainties that could have the greatest impact, by considering both the range of likelihoods and potential consequences. For example, the impact of a tropical storm largely depends on when and where it makes landfall, while a heat wave will have different impacts on different populations depending on their social vulnerabilities. Given that better information will become available over time, and new risks will become apparent, the analysis of climate risks needs to be an iterative process that monitors the evolution of the characteristics of risks over time. By emphasising flexibility and learning, a risk-based approach can help to respond to changing circumstances and new information.

Notes

1. These projections do not fully capture the impact of changes in extreme weather events, increased water stress or the loss of ecosystem services and assets.
2. Disasters in the EM-DAT database include events where: ten or more people were killed; 100 or more people were affected, injured or made homeless; significant damage was incurred; a declaration of a state of emergency and/or an appeal for international assistance was made. Economic losses are defined as the direct (e.g. damage to infrastructure, crops, housing) and indirect (e.g. loss of revenues, unemployment, market destabilisation) consequences of a disaster on the local economy.

References

AON Benfield (2014), *Annual Global Climate and Catastrophe Report: Impact Forecasting 2013*, Impact Forecasting, Chicago.

ClimateCost (2011), The Costs and Benefits of Adaptation Policy in Europe: Review Summary and Synthesis, ClimateCost Policy Brief, *www.climatecost.cc/images/Review_of_European_Costs_and_Benefits_of_Adaptation.pdf* (accessed 2 December 2014).

Climate Council of Australia (2014a), *Counting the Costs: Climate Change and Coastal Flooding*, Climate Council of Australia, Potts Point.

DCCEE (2011), "Climate Change Risks To Coastal Buildings And Infrastructure: A Supplement To The First Pass National Assessment", Department of Climate Change and Energy Efficiency, Canberra.

Dow, K. et al. (2013), "Limits to adaptation", *Nature Climate Change*, Vol. 3, *http://dx.doi.org/10.1038/nclimate1847*.

EM-DAT (*Emergency Event Database*) (n.d.), "The International Disaster Database", Centre for Research on the Epidemiology of Disasters, *www.emdat.be/* (accessed 27 February 2015).

Griffin, D. and K. Anchukaitis (2014), "How unusual is the 2012-2014 California drought?", *Geophysical Research Letters*, Vol. 41(24), *http://dx.doi.org/10.1002/2014GL062433*.

Howitt, R.E. et al (2014), *Economic Analysis of the 2014 Drought for California Agriculture*, Center for Watershed Sciences, University of California, Davis, California.

IPCC (Intergovernmental Panel on Climate Change)(2014a), *Climate Change 2014: Impacts, Adaptation, and Vulnerability*, Contribution of Working Group II to the Fifth Assessment Report of the Intergovernmental Panel on Climate Change, Cambridge University Press, Cambridge and New York.

IPCC (2014b), *Climate Change 2014: Mitigation of Climate Change*, Contribution of Working Group I to the Fifth Assessment Report of the Intergovernmental Panel on Climate Change, Cambridge University Press, Cambridge and New York.

IPCC (2013), *Climate Change 2013: The Physical Science Basis*, Contribution of Working Group I to the Fifth Assessment Report of the Intergovernmental Panel on Climate Change, Cambridge University Press, Cambridge and New York.

IPCC (2012), *Managing the Risks of Extreme Events and Disasters to Advance Climate Change Adaptation*, A Special Report of Working Groups I and II of the Intergovernmental Panel on Climate Change, Cambridge University Press, Cambridge and New York.

IPCC (2007), *Climate Change 2007: Synthesis Report*, Contribution of Working Group I, II and III to the Fourth Assessment Report to the Intergovernmental Panel on Climate Change, Cambridge University Press, Cambridge and New York.

Jongman, B. et al. (2014), "Increasing Stress on Disaster-Risk Finance due to Large Floods", *Nature Climate Change*, Vol. 4, *http://dx.doi.org/10.1038/nclimate2124*.

Kuczinski, T. and K. Irvin (2012), "Severe weather in North America: Perils Risks Insurance", *Knowledge Series, Natural Hazards*, Munich Re. Munich.

Liberato, M.L.R. et al. (2013), "Explosive development of winter storm Xynthia over the subtropical North Atlantic Ocean", *National hazards and Earth Systems Sciences*, No. 13, *http://dx.doi.org/10.5194/nhess-13-2239-2013*.

Mechler, M. et al. (2014), "Managing unnatural disaster risk from climate extremes", *Nature Climate Change*, No. 4, *http://dx.doi.org/10.1038/nclimate2137*.

Munich Re (2015), "Loss events worldwide 2014: Percentage distribution", Annual statistics 2014, *www.munichre.com/en/reinsurance/business/non-life/natcatservice/annual-statistics/index.html* (accessed 27 February 2015).

Munich Re (2006), *The 1906 earthquake and Hurricane Katrina: Similarities and differences – Implications for the insurance industry*, Münchener Rückversicherungs-Gesellschaft, Munich.

OECD (Organisation for Economic Co-operation and Development) (2015 forthcoming), *The economic consequences of climate change*.

OECD (2014), *Boosting Resilience through Innovative Risk Governance*, OECD Publishing, Paris, *http://dx.doi.org/10.1787/9789264209114-en*.

OECD (2012), *OECD Environmental Outlook to 2050: The Consequences of Inaction*, OECD Publishing, Paris, *http://dx.doi.org/10.1787/9789264122246-en*.

Swedish Commission on Climate and Vulnerability (2007), *Sweden Facing Climate Change – Threats and Opportunities*, Swedish Government Official Reports SOU 2007:60, Stockholm.

The ice2sea Consortium (2013), *From Ice to High Seas: Sea-level rise and European coastlines*, The ice2sea Consortium, Cambridge, United Kingdom.

van Bree. L, and van der Sluijs. J (2014), "Background on Uncertainty Assessment Supporting Climate Adaptation Decision-Making", in Lourenço, T.C., et al. (eds.), *Adapting to an Uncertain Climate*, Springer, Switzerland.

Climate Change Risks and Adaptation
Linking Policy and Economics

Chapter 2

Approaches to managing climate risks

This chapter provides an overview of the approaches used by OECD countries to manage their exposure to climate risks. It highlights: the need to combine political will; effective institutional structures; and tools and evidence to ensure an effective response to climate change. It explores how national governments can support activity by local governments and the private sector. It emphasises the importance of measuring progress in adaptation using monitoring and evaluation.

Key messages

- Integrating adaptation into decision making requires aligning political will, strengthening institutional arrangements and applying appropriate economic tools.
- National governments have a crucial role to play in supporting adaptation by households, businesses and local administrations. Key areas for achieving this include: providing access to information, tools and guidance; ensuring that regulations are coherent and avoid perverse incentives; addressing social drivers of vulnerability; and considering climate risks when making funding decisions.
- Monitoring and evaluation will be essential to assess the effectiveness of countries' approaches to adaptation, identify emerging risks and respond to changing policy contexts. National strategies for adaptation have been adopted by 24 OECD countries, but the development of strategies for measuring progress remains at an early stage for most of them.

National measures to address climate change risks

The development of national adaptation strategies is a common feature of OECD countries' preparations for climate change. In 2005, Finland was one of the first countries to publish a national adaptation strategy. Currently, 24 OECD countries have published national strategies and 7 are in the process of developing them (see Table 2.1). A common objective of these strategies is to demonstrate political commitment on adaptation, communicate the government's overall approach to adaptation, and facilitate co-ordination. The strategies also help to improve the evidence base on climate change impacts and vulnerabilities, and to identify areas where there is scope to enhance the country's adaptive capacity and enabling environment (Mullan et al., 2013; Gagnon-Lebrun and Agrawala, 2006).

The strategies vary in their prescriptiveness. For example, Norway's 2013 White Paper on Climate Change Adaptation states that individuals, the private and the public sectors alike have an obligation to adapt to climate change, but that it is the government's responsibility to put in place an enabling environment that facilitates that process (Climate-Adapt, n.d.). Other countries have complemented their strategic adaptation objectives with more detailed guidance on how they can be achieved. For example, France's 2006 national adaptation strategy was complemented with a national adaptation plan in 2011, which outlines 84 specific adaptation actions expressed in 230 measures (French Government, 2011). This is also the approach taken by Germany where an action plan in 2011 was published to complement the 2008 adaptation strategy (BMU, 2011; 2008).

Some OECD countries are implementing adaptation at the national level without adopting a dedicated strategy or plan. For example, Japan published in 2010 the Approaches to Climate Change Adaptation guidance, targeting policy makers responsible for designing and evaluating adaptation responses to identified climate risks (Mullan et al., 2013). Similarly, Canada introduced in 2007 a four-year investment programme (CAD 85.9 million) to encourage and support climate change action by provinces, territories, municipalities

Table 2.1. **Coverage of adaptation in National Communications**

		Assessment of climate data			Adaptation options and policy responses				
		Historical climatic trends	Climate change scenarios	Impact assessments	Identification of adaptation options	Mention of policies synergistic with adaptation	Establishment of institutional mechanisms for adaptation responses	Formulation of adaptation policies	Explicit incorporation of adaptation in projects
No adaptation strategy published	Canada	●*	○	●	●	○	●*	●	●*
	Czech Republic – planned for 2016	●	●	●	●	○		●*	●*
	Estonia – planned for 2016	●*	●*	●	○	○	○	○*	
	Greece – under development	●	●	●	●	●	○*	○	○*
	Iceland	●	●*	●					
	Israel – under development	○	○	●	●	○	○	●	
	Italy – under development	●	○	●	●	●		○	○
	Japan – planned for 2015	●*	●	●	●	○		○	
	New Zealand	●*	●	●	●	●*	○	●	
	Slovenia – under development	○*	○	●	○			○	
Adaptation strategy published	Australia	○	●	●	●	●*	●	●	●*
	Austria	●*	●	●	●*	●	○*	●*	○
	Belgium	●*	●	●	●	●	●*	●	
	Chile	○	●	●	●	○	○*	○	
	Denmark	●	●*	●	●*	●*	○	●	●
	Finland	○*	●*	●	●	●	○	●	○
	France	●	●	●	○	●	●*	●	
	Germany	○	●*	●	○	○	○*	●	○
	Hungary	○*	●	●*	●*	○*	○*	○	○
	Ireland	●	●*	●	●*	●	●*	○	
	Korea	●	●	●	●	●		●	
	Luxembourg	●*	○	●*	●*	●*		●*	
	Mexico	●	●	●	●	●	○	●	
	Netherlands (new NAS in 2016)		●*	●	●	○	○*	●	●*
	Norway	●	●	●	●*	●*	●	●*	
	Poland	○	●	●	○	●*	○*	●*	
	Portugal	○*	○*	●*	●*	●*	●*	●	
	Slovak Republic	●	●	●	●	●*	*	●*	○*
	Spain	○	●	●	●	●*	●*	●	
	Sweden	○	●	●	●	○	●*	○	
	Switzerland	●	●	●	●*	●	●	●*	●
	Turkey	●	●	●	●*	○*	○*	●*	●*
	United Kingdom	○	●*	●	●	●	●	●	●*
	United States	●	●	●	●	●*	●*	●*	●*

Coverage in NCs:

- Extensive discussion
- Some mention/limited discussion
- No mention of discussion

* Changes that occurred since last National Communication published

Quality of discussion in NCs:

● Discussed in detail, i.e. for more than one sector or ecosystem, and/or providing examples of policies implemented, and/or based on sectoral/national scenarios

○ Discussed in generic terms, i.e. based on IPCC or regional assessments, and/or providing limited details/no examples/only examples of planned measures as opposed to measures implemented

Note: The statistical data for Israel are supplied by and under the responsibility of the relevant Israeli authorities. The use of such data by the OECD is without prejudice to the status of the Golan Heights, East Jerusalem and Israeli settlements in the West Bank under the terms of international law.

Source: Updated from Mullan, M. et al. (2013), *National Adaptation Planning: Lessons from OECD Countries*, OECD Environment Working Papers, No. 54, OECD Publishing, Paris, *http://dx.doi.org/10.1787/5k483jpfpsq1-en*; Gagnon-Lebrun, F. and S. Agrawala (2006), *Progress on Adaptation to Climate Change in Developed Countries: An Analysis of Broad Trends*, OECD Publishing, Paris.

and professional organisations through enhanced climate change knowledge and capacity. In 2011, an additional CAD 148.8 million was allocated to extend the programme to 2016. Alongside this, a Federal Adaptation Policy Framework was developed to help mainstream climate change into federal decision making (Government of Canada, 2014).

Countries efforts to manage climate risks have targeted three interlinked aspects of the policy making process (Persson and Klein, 2009): i) political commitment to address climate change risks; ii) the introduction of institutional frameworks and organisational processes that enable the implementation of policies required to address climate change risks and opportunities; and iii) the development of tools and evidence to guide decision-making processes.

Political commitment to address climate change risks

Political commitment is essential to plan and implement adaptation actions and to mobilise the resources needed. Securing this commitment is particularly important given the long-term, complex and uncertain nature of climate change. Political leadership is also required to overcome institutional inertia and change long-established approaches to policy development (Persson and Klein, 2009).

Extreme climate events can generate political will to address longer-term climate risks. Although it is not possible to conclusively attribute specific events to climate change, their consequences can expose underlying vulnerabilities and focus attention on the need for them to be addressed (IPCC, 2007). For instance, President Barack Obama's Climate Action Plan includes a pillar on the need to better prepare the United States for the impacts of climate change. This includes a programme focused on "Building Stronger and Safer Communities and Infrastructure" centered around "(r)ebuilding and (l)earning from Hurricane Sandy" (Executive Office of the President, 2013) (see Box 2.1). Similarly, the 2003 European heat wave estimated to have led to 35 000 excess deaths across Europe (IPCC, 2007), contributed to the implementation of the French heat wave plan ("*Plan canicule*") (see Box 2.2).

Box 2.1. **The US President's Climate Action Plan**

President Obama's 2013 Climate Action Plan consists of three pillars: i) to reduce carbon emissions; ii) to prepare for the impacts of climate change; and iii) to demonstrate global leadership on climate change. The second pillar consists of three interrelated initiatives:

- **Building Stronger and Safer Communities and Infrastructure:** This initiative supports states, cities and communities in their efforts to enhance the resilience of their infrastructure to climate change. Federal agencies are encouraged to make climate-resilient investments and to review counter-productive policies. A task force will also be established to advise on how the federal government can best support local preparedness and resilience building efforts. Targeted support will be provided to communities to e.g. enhance the resilience of the transportation, buildings and infrastructure sectors, to assist the most vulnerable communities, and to rebuild areas affected by Hurricane Sandy.
- **Protecting our Economy and Natural Resources:** Recognising that climate change will affect every aspect of society, this initiative explores how vulnerable key sectors are to climate change. It will identify efforts that can protect these sectors while at the same time targeting hazards that cut across sectors and regions. This will inform the implementation of sector- and hazard-specific efforts in areas of health, insurance, land and water management, agriculture, drought, wildfire and floods.

Box 2.1. **The US President's Climate Action Plan** *(cont.)*

- **Using Sound Science to Manage Climate Impacts:** To help decision makers better understand and manage climate risks, this initiative helps to advance the climate science and the development of tools that can inform climate-relevant decisions. To achieve this objective, the availability, accessibility, and utility of relevant scientific tools and information will be enhanced.

The Climate Action Plan complements a number of measures already in place, such as the Interagency Climate Change Adaptation Task Force established in 2009, the Executive Order (EO) 13514 on Federal Leadership in Environmental, Energy, and Economic Performance (including climate change adaptation and mitigation) issued in 2009, and the decadal National Global Change Research Plan implemented in 2012.

Source: Executive Office of the President (2013), *The President's Climate Action Plan*, The White House, Washington, DC.

Box 2.2. **Preparing for heat waves: The French "*Plan canicule*"**

The 2003 Europe heat wave is estimated to have contributed to around 15 000 excess deaths in France, predominantly among the elderly. This was the first time France faced the potential consequences of extreme heat and it revealed the need for a national heat wave plan – "Plan canicule" – to be adapted for each "département" or county. The objective of the plan is to anticipate the health effects of heat waves and to alert authorities in a timely manner to allow them to set up preventative action. The plan includes four levels:

- **Level 1 (green alert):** No alert in place.
- **Level 2 (yellow alert):** Minor alert – heat wave forecast for three days. Prevention and protection to be put into effect.
- **Level 3 (orange alert):** High alert – effective heat wave. People are encouraged to act accordingly.
- **Level 4 (red alert):** Maximum mobilisation – heat wave with severe medical impact expected.

During the summer, weather reports on TV, radio and in newspapers quote the four levels of alert. This can help to inform peoples' decision-making processes, while the different levels also activate responses by those responsible for public health surveillance, social support and medical preventative action.

Source: Pirard, P. et al. (2005), *Summary of the Mortality Impact Assessment of the 2003 Heat Wave in France*, Eurosurveillance, Vol. 10, Issue 7.

The European Union is increasingly active in supporting adaptation efforts by its member states, underpinned by the adoption of the EU adaptation strategy in 2013 (Council of the European Union, 2013). The strategy encourages, but does not yet require, member states to develop national adaptation strategies. A self-assessment survey undertaken by the European Environment Agency (2014) found that 28 out of the 30 countries that answered the survey identified extreme weather events as one of the three most important triggers for national action on adaptation. Other triggers commonly mentioned were: development of EU policies (identified by 19 out of 30 respondents); estimates of current and future damage (17 of 30); and pertinent results from scientific research (14 of 30).

Civil society, parliamentarians and businesses can play a key role in mobilising and sustaining political commitment. They played an important role in galvanising support for the Climate Change (Scotland) Act 2009 (Wolstenholme, 2010). Some parts of the private sector have also been active in urging for increased action on adaptation. In particular, the insurance industry has called for increased investment in research and advocating policy reforms to reduce underlying risks.

Designing adaptation measures that deliver near-term benefits can help to generate and sustain political commitment when the benefits are realised over the near-term. However, decisions with long-term implications, such as infrastructure development and spatial planning, do not necessarily have these characteristics. Managing those risks can entail incurring (financial and opportunity) costs now for future (and potentially uncertain) benefits. In practice, adaptation strategies have primarily focussed on identifying methods to address current risks in ways that can also yield longer-term benefits (Mullan et al, 2013).

Institutions and processes to address climate change risks

Political will needs to be combined with institutional arrangements that enable climate risks to be considered at the right points in the decision-making process (Persson and Klein, 2009). Common practice in OECD countries has been to mainstream adaptation within existing ministerial responsibilities. All government ministries and agencies are encouraged to consider the possible impacts of climate change and, when necessary, to include additional resilience measures in their planning and budget processes. Examples of such mainstreaming measures include revisions to government regulations (e.g. building standards) and integration of climate risks during the appraisal of projects or programmes (see Chapter 6). For instance, in 2013 all US agencies were required to include Climate Change Adaptation Plans in their Sustainability Plans, identifying and assessing the potential impacts from climate change on the agency's ability to accomplish its missions, operations and programmes (US Department of State, 2014). Similarly, in Germany, Federal Government agencies are required to consider climate change when developing legal provisions and encouraged to integrate it into standards, regulations and into government funding programmes where appropriate (see Box 2.3).

Box 2.3. **The German institutional framework to address climate change risks**

In Germany, the Adaptation Action Plan introduced a number of regulations to ensure that climate change is considered in national planning processes. These regulatory measures include:

- Legal provisions: All federal ministries are required to examine whether it is possible and appropriate to consider climate change risks and adaptation requirements as targets, principles or possible trade-offs in relevant legislations.
- Standards and regulations: All government bodies responsible for setting standards and regulations are required to examine if and how climate change considerations can be included in these processes. This can support enterprises in their decision-making processes, at the same time as it provides a level of legal security if the revised standards and regulations reflect generally accepted recommendations.

Box 2.3. The German institutional framework to address climate change risks *(cont.)*

- Funding programmes: All new government funded initiatives are encouraged to consider climate change risks. In addition to this, the Federal Government will examine the extent to which existing programmes, both at the national and at the EU level, address adaptation to climate change.

Source: BMU (2011), *Adaptation Action Plan for the German Strategy for Adaptation to Climate Change*, Berlin, BMU.

Weak co-ordination between ministries and agencies working on linked policy objectives (e.g. climate change, disaster risk reduction and natural resources management) can impede planning and implementation of adaptation. This can result in duplicative or conflicting adaptation efforts being implemented across the government. Alternatively, efforts to achieve adaptation objectives might dilute other important sectoral objectives (e.g. water quality targets) (Brouwer, Rayner and Huitema, 2013). Across OECD countries, the responsibility for co-ordinating action on adaptation is usually located in environment ministries, rather than a central ministry or executive office.

Tools and evidence to inform a national response to climate change risks

Political will and institutional arrangements provide the impetus to consider climate risks, but tools and evidence are required to develop an appropriate response to those risks. The development of tools and evidence has been a major focus of adaptation efforts in OECD countries. For example, the German Competence Centre for Climate Impacts and Adaptation (KomPass) provides ongoing support for adaptation activities undertaken by both public and private sector actors.

Investment in climate research has contributed to improved projections of climate change that include higher levels of spatial detail and more explicit treatment of some sources of uncertainty. To make informed decisions on adaptation, policy makers must actively engage with this research and base planning and budgeting processes on targeted and robust climate information. However, a number of bottlenecks may prevent adaptation research from being translated into adaptation measures (Klein and Juhola, 2013):

- Theoretical adaptation concepts do not always relate to the "reality" facing decision makers.
- The uncertain nature of some potential impacts makes stakeholders inclined to delay action, rather than to act now.
- Divergences between the scale at which decision makers operate and the scale of information provided by climate information models.
- A mismatch between stakeholders' primary concern to manage current climate variability and the longer-term focus of much existing adaptation research.

The role of sub-national governments in addressing climate risks

The consequences of climate change are experienced at the local and regional levels, and many of the policy levers for managing those risks are held by local and regional governments. These levers include the enforcement of building codes, planning decisions and the provision of emergency services. Urban areas are particularly vulnerable to the impacts of climate change due to their high population density and aging infrastructure, among other factors. Around two-thirds of people in OECD countries live in urban areas, which can be vulnerable to climate change risks such as heat waves, flooding, water scarcity and storm surges. An example of the scale of potential impacts was provided by the 2002 river flooding in Dresden, Germany. This flood led to 28 deaths and caused damages on infrastructure, monuments and buildings estimated at around EUR 9.2 billion (ICLEI and CEPS, 2013).

Local governments are increasingly supporting and driving adaptation initiatives within the framework provided by national or state level legislation. Many of them have started to assess their climate risks and vulnerabilities and to identify suitable adaptation measures (see Table 2.2 some examples). Chicago's assessment of its climate risks and vulnerabilities in 2008 informed the city's Climate Action Plan. The Plan builds upon existing greening and sustainability initiatives and is structured around five strategic areas: i) energy efficient buildings; ii) clean and renewable energy sources; iii) improved transportation options; iv) reduced waste and industrial pollution; and v) adaptation. The adaptation component includes innovative approaches to cooling, the development of a watershed plan, and the implementation of Chicago's Green Urban Design Plan to manage heat and flooding (City of Chicago, 2008).

Table 2.2. **Examples of local level approaches to adaptation planning**

	Type of citywide assessment		Type of local adaptation plan			Approach to integration	
	Hazard assessment	Vulnerability assessment	Include adaptation in local climate action plan	Create strategic adaptation plan	Develop dedicated sector plan(s)	Mainstream into policies or sectors	Incorporate into sector plan(s)
Amman (Jordan)	Intended	Intended	Intended			Intended	
Boston (US)			√			√	
Copenhagen(Denmark)	√	√	√	√	Intended	Intended	Intended
Durban (South Africa)	√			√	√		√
London (UK)	√	√		√			
Maputo (Mozambique)	√						
Quito (Ecuador)	Intended	Intended	√	Intended	Intended	√	Intended
Seattle (US)	√	√	√			√	√
Semarang (Indonesia)	√	Intended	Intended	√		Intended	
Surat (India)	√		√	√	√	√	Intended
Taipei (Taiwan)							
Tokyo (Japan)	√						
Toronto (Canada)	√	√	√	√	Intended		√
Walvis Bay (Namibia)	√			Intended		√	

Source: Carmin, J., D. Dodman and E. Chu (2013), *Urban Climate Adaptation and Leadership: From Conceptual Understanding to Practical Action*, OECD Regional Development Working Papers, No. 2013/26, OECD Publishing, Paris, *http://dx.doi.org/10.1787/5k3ttg88w8hh-en.*

Some cities have introduced adaptation measures in advance of national strategies or plans being developed. Tokyo, for example, has constructed super levees (typically 10 metre high and more than 200 metre wide) designed to withstand extreme weather events and resistant to overflow, seepage and earthquakes (Stalenberg, 2012). The super levees have been integrated into the development of parks and other recreational facilities. Similarly, the Tokyo Metropolitan Government and the city of Yokohama have jointly developed multi-purpose storage basins by the Tsurumi River to protect against 150-year floods. The catchment basin, with a storage capacity of 3.9 million cubic metres, has been transformed into a recreational area with parks and sports facilities (Stalenberg and Kikumori, 2008).

However, the support of national governments is recognised as an important factor contributing to the success of local initiatives (OECD, 2009). For instance, in New Zealand, municipal adaptation initiatives have benefitted from support by the national government for the provision of reliable and relevant climate data, the translation of national level risks to local level vulnerabilities, and the facilitation of political buy-in for local adaptation actions (Hunter, Burkitt and Trangmar, 2010).

As providers of important public services (e.g. health care and emergency services) and infrastructure, local governments also play a critical role in ensuring that their policies and investment decisions adequately reflect climate change risks. The government of Denmark has encouraged municipalities to draft local adaptation plans to complement the country's 2008 National Adaptation Strategy and the 2012 Action Plan. To facilitate this process, a Climate Change Adaptation Squad and a mobile task force were established. Several legislative changes have also been introduced that enable municipalities to cite adaptation as a basis for local plans (Goodsite et al., 2013) (Box 2.4).

Box 2.4. **Copenhagen's Climate Adaptation Plan**

The City of Copenhagen adopted its Climate Adaptation Plan in 2011. One of the priorities for this plan is to address the risks posed by cloudbursts on the city's sewage system. Currently, the acceptable level of risk is that wastewater should not reach ground level more than once every 10 years. While this will remain the same, the action plan specifies that average water levels must not rise more than 10 centimetres above ground level more than once every 100 years, with the exception of places specifically designated for storage of surface water. To achieve this goal, the plan identifies DKK 3.8 billion of measures that will both make the city greener and keep it blue by diverting the water over ground when possible. When this is not possible, for example in the case of densely populated areas, underground tunnels will be used. The implementation of measures in this plan is prioritised on the basis of: targeting areas at high risk; ease of implementation; and synergies with related initiatives.

Source: City of Copenhagen (2012), *Københavns Kommunes Skybrudsplan 2012*, City of Copenhagen, Copenhagen.

Support for mainstreaming adaptation in local areas should be shaped by the level of capacity in those areas and the organisation of responsibilities. There can be fewer people and resources to implement adaptation at the local level than at the national level, particularly for smaller or less wealthy localities. However, the administrative structures of local governments can favour better integrated decision making, as a team, or even a staff member, is often engaged in multiple policy agendas at the same time. The consequences of this are that tools and evidence are more likely to be used if they can support the achievement of multiple objectives (e.g. climate adaptation and mitigation, land use management, and infrastructure) and that there is a particular premium on them being easy to use.

Private sector engagement in adaptation

Progress on adaptation will depend upon the decisions made by the private sector and these decisions will primarily be driven by the profit motive. Adaptation measures can reduce costs of disruption to business operations (e.g. increasing costs of maintenance and materials, and raising insurance costs) and help to realise new business opportunities arising from climate change (Agrawala et al., 2011). An example of the potential scale of risks was provided during the 2003 heatwave. This event cost a French energy company

(*Electricité de France*) USD 431 million, as a result of having to limit and suspend the operation of several power plants. This was due to regulatory restrictions on the temperature of discharge water, as well as the high cost of having to purchase electricity on the open market (Stenek, Amado and Greenall, 2013). As well as risk management, business opportunities could include the expansion of markets for new and innovative products for resource efficiency or participation in publicly funded adaptation projects (Pauw and Pegels, 2013).

OECD analysis of private sector engagement found that the majority of companies surveyed were aware of climate risks, but few were taking action to address those risks (Agrawala et al, 2011). Out of the 1 100 English language responses to the 2009 Carbon Disclosure Project (CDP), 75% of respondents stated that they had considered the potential impact of climate risks to their businesses. The majority of those companies (59%) did not take any further action in response to those risks. Only 7% of the companies aware of climate risks stated that they had identified possible adaptation responses, such as improved business practices to investments in infrastructures and technologies (Agrawala et al., 2011).

There are several reasons why activity has lagged behind awareness. Visible adaptation initiatives may also understate the actual level of adaptation taking place since many companies do not distinguish adaptation initiatives from their standard risk management processes (Agrawala et al., 2011). Companies can also be cautious about disclosing information on their adaptation measures for competitiveness reasons. In some cases, a decision not to adapt based on the company's self-interest may be a rational response to the company's operating contexts (e.g. supply chain flexibility) and the long time horizon of climate change (Agrawala et al., 2011). Market failures and policy distortions can also hinder action, which is particularly problematic for sectors that are economically significant and where decisions have long-term implications.

National governments can facilitate private sector management of climate risks. As an initial step, this may entail providing access to information and raising awareness on climate change risks and opportunities. Additional measures introduced by OECD governments to encourage and enable private sector action include (Mullan et al., 2013):

- **Providing tools and guidance:** High-quality and accessible climate data, guidance on managing climate issues, and tools to assessing climate risks and planning adaptation responses.
- **Maintaining regulatory coherence:** Removing distorting incentives and ensuring that regulatory frameworks that are conducive to risk management, including the management of overlapping regulatory regimes (e.g. regulation of water, energy and land use).
- **Establishing reporting requirements:** Mechanisms to ensure that companies in key sectors (e.g. providers of critical national infrastructure) regularly assess and report their exposure to climate risks.
- **Using procurement policies**: Governments have used their procurement policies to encourage and require private sector suppliers to take account of climate risks, while considering the importance of non-discriminatory competition policies for both domestic and imported products.

These four measures overlap with the five areas that have been identified by the International Finance Corporation (Stenek, Amado and Greenall, 2013) as important in creating an enabling environment for private sector adaptation: i) data and information; ii) institutional arrangements; iii) policies; iv) economic incentives; and v) communication, technology and knowledge (see examples of each category in Table 2.3).

Table 2.3. **Factors influencing climate change adaptation in the private sector**

Data and information	Example
• Access to climate and hydrological observations and/or projections. • Data and information readily available on selected impacts, taking into account climate projections. • Decision-support tools to understand and assess risks and opportunities, and/or identify and select adaptation actions.	To bridge the gap between climate science and adaptation, CLIM Systems Ltd has developed a software modelling system called SimCLIM that can help assess climate impacts and inform adaptation to climate variability and change. The software is global in scope and the methods applied are consistent with IPCC guidelines.
Institutional arrangement	**Example**
• Co-ordinating agencies (e.g. governments, private sector, civil society, NGOs and academia) that have activities focused on climate risk and adaptation. • Public-private partnerships dedicated to assessing and tackling climate change adaptation challenges. • Government and/or industry organisations that provide support to alternative productions/activities and/or relocation in the private sector.	Research4Life is an academic public-private partnership that grants students, researchers and academics access to online scientific journals related to climate related studies. Eye On Earth is a partnership with the European Environment Agency that helps to deliver real-time environmental information to citizens across the globe.
Policies	**Example**
• Building codes and standards taking into account changing climate conditions and the associated impacts on building design and operations. • Local zoning regulations incorporating data/information about future changes in the climate and in the projected impacts on new and/or existing infrastructure and buildings. • Land use/construction permitting rules promoting climate change adaptation measures (e.g. permits used to promote tree planting to cool urban areas or to absorb more water where the Urban Heat Island effect or flooding pose risks).	URS Scott Wilson was commissioned by the 3 Counties Alliance Partnership (3CAP) in the United Kingdom (i.e. Nottinghamshire, Derbyshire and Leicestershire) to carry out a collaborative study to investigate the impact of climate change on their highway policies and standards and to identify adaptation opportunities.
Economic incentives	**Example**
• Incentives in support of purchases of climate change adaptation technologies and/or implementation of adaptation actions and/or R&D in the private sector. • Public and private funding instruments in support of climate change adaptation uptake in the private sector. • Charges and/or levies used to fund climate change adaptation works in critical public infrastructure.	The mayor of Boston, Thomas Menino, issued an Executive Order on Climate Action in 2007. The Order mandates that all municipal construction projects and major renovations evaluate the climate change risks and identify steps to avoid, minimise or reduce those risks.
Technology and knowledge	**Example**
• Climate change adaptation technologies and/or process innovation are produced, sold and/or promoted in the private sector. • Professional post-secondary education curriculums incorporate climate change impacts and adaptation knowledge and/or training.	Bayer Crop Science carries out research on plants more resistant to climatic conditions and improves their stress tolerance. The company's Stress Shield products, for example, have already successfully established themselves on the market.

Source: Based on Stenek, V., J.C. Amado and D. Greenall (2013), *Enabling Environment for Private Sector Adaptation: An Index Assessment*, International Finance Corporation, Washington DC; UNFCCC (n.d.), *Nairobi Work Programme Private Sector Initiative – database of actions on adaptation*; Jacobs et al. (2011) *Legal options for municipal climate adaptation in South Boston.*

Addressing the social aspect of adaptation

People's vulnerability to climate change is closely linked to drivers of social vulnerability, such as poverty and social exclusion. Some segments of the population (e.g. the elderly, displaced persons, or people with mobility challenges, poor health or low income) are likely to be disproportionately affected by emerging climate change risks (Boeckmann and Zeed, 2014; Council of the European Union, 2013). People's ability to adapt to climate risks depends upon resource availability, access to information and technology, and the presence of institutions and infrastructure (Bauriedl, 2011). Understanding the diversity of interests, values and socio-economic contexts can improve the effectiveness of adaptation policies (IPCC, 2014).

The treatment of social aspects varies across OECD countries, with only a few strategies including an explicit objective on enhancing the resilience of the most vulnerable groups. However, many national adaptation strategies make reference to the social implications of climate change in the context of spatial exposure, such as communities located in flood prone areas. Another common focus lies with the health implications of hotter weather, particularly on the older population. Less attention has been paid to the wider social dimensions of vulnerability, such as the ability of individuals and groups to prepare for, respond to, and recover from climate impacts, taking into account social factors, such as income and the nature of people's social networks (Brisley et al., 2012).

Considering both climate and social vulnerability has the potential to encourage a more inclusive policy process. At present climate change policy and policies seeking to tackle social vulnerability and poverty are largely developed and implemented in isolation. Recognising the social dimensions of vulnerability to climate change can support an enhanced dialogue between these policy areas. This can entail a broadening of focus beyond the areas traditionally viewed as adaptation. Policies that are concerned with the care of the elderly, the quality of neighbourhoods and levels of material deprivation are all important for climate adaptation (Lindley et al., 2011). "Adaptation-proofing" the policies developed by different governmental agencies led by, for instance the health, transport and agriculture ministries can also be useful to account for diverse needs and circumstances.

The importance of measuring progress in addressing climate risks

As countries implement national approaches on adaptation, it is essential to monitor and evaluate whether they are succeeding in becoming more resilient to climate change. At the national level, however, interventions motivated by climate change are only one driver of resilience. Other factors such as socio-economic trends and non-climate policies also play an important role (OECD, 2015). Monitoring and evaluation provides a set of tools that can be used to assess the effectiveness of different approaches to adaptation.

The appropriate design of a national monitoring and evaluation framework will depend on countries' approaches to adaptation. For example, when countries take a mainstreamed approach and no additional resources are allocated for adaptation, the implementation of the adaptation component of specific initiatives cannot be monitored; nor can their cost-effectiveness be assessed. Instead, monitoring and evaluation can shed light on changes in identified adaptation priorities and in the country's overall vulnerability to climate change. Alternatively, when adaptation plans outline a set of specific adaptation initiatives to be implemented within a given timeframe, the monitoring and evaluation framework will focus more on whether planned initiatives have been implemented and if they deliver agreed objectives (OECD, 2015).

Although the majority of OECD countries are still in the early phase of developing monitoring and evaluation frameworks for adaptation, an emerging trend has been for countries to initially monitor prioritised climate risks identified through climate risk and vulnerability assessments. These assessments are primarily a tool intended to identify priority risks and inform planning, but they can also provide a good basis against which subsequent changes in adaptation priorities can be assessed (OECD, 2015). If repeated on a regular basis, as is for example the case in Norway and in the United Kingdom, the assessments can provide periodic "snapshots" of the adaptation priorities and the emerging priority risks and vulnerabilities. To complement these periodic assessments, indicators can be developed to monitor the evolution of risks over time (OECD, 2015).

In addition to monitoring specific risks, changes in countries' overall vulnerability or resilience to climate change must also be assessed. France's monitoring and evaluation framework, for example, combines regular monitoring of adaptation activities outlined in the national adaptation plan with targeted evaluation of key sectors using a range of evaluation techniques, such as impact assessment, cost-effectiveness and cost-benefit analysis. Lessons learned from specific projects or programmes can be used to complement the collection of aggregate information. Such evaluations can, for example, focus on initiatives that pilot innovative approaches to adaptation. Further, national audits can assess if public expenditure on adaptation is aligned with international and national policy goals, and spent in accordance with existing rules, regulations and principles of good governance (OECD, 2015).

The insights generated from monitoring should be used to inform planning and budgeting processes. To achieve this objective, it is important that the information is accessible and made available in a timely manner. For example, the publication of risk assessments or evaluations can coincide with national planning and budgeting processes. It may also be necessary to change the incentive structure of public officials, and institutional arrangements, to ensure that the information gathered informs subsequent planning and budgeting processes (Mackay, 2007).

References

Agrawala, S., et al. (2011), "Private Sector Engagement in Adaptation to Climate Change: Approaches to Managing Climate Risks", *OECD Environment Working Papers*, No. 39, OECD Publishing, Paris, *http://dx.doi.org/10.1787/5kg221jkf1g7-en*.

Bauriedl, S. (2011), "Adaptive Capacities of European City Regions in Climate Change: On the Importance of Governance Innovations for Regional Climate Policies", in Filho, W.L. (ed.), *The Economic, Social and Political Elements of Climate Change*, Springer-Verlag, Berlin.

BMU (2011), *Adaptation Action Plan for the German Strategy for Adaptation to Climate Change*, The German Federal Ministry for the Environment, Nature Conservation and Nuclear Safety (BMU), Berlin.

BMU (2008), *German Strategy for Adaptation to Climate Change*, The German Federal Ministry for the Environment, Nature Conservation and Nuclear Safety (BMU), Berlin.

Boeckmann, M. and H. Zeeb (2014), "Using Social Justice and Health Framework to Assess European Climate Change Adaptation Strategies", *International Journal of Environmental Research and Public Health*, Vol. 11(12), *http://dx.doi.org/10.3390/ijerph111212389*.

Brisley et al. (2012), *Socially Just Adaptation to Climate Change*, Joseph Rowntree Foundation, York.

Brouwer, S., T. Rayner and D. Huitema (2013), "Mainstreaming climate policy: the case of climate adaptation and the implementation of EU water policy", *Environment and Planning C: Government and Policy*, 31(1), *http://dx.doi.org/10.1068/c11134*.

Carmin, J., D. Dodman and E. Chu (2013), "Urban Climate Adaptation and Leadership: From Conceptual Understanding to Practical Action", *OECD Regional Development Working Papers*, No. 2013/26, OECD Publishing, Paris, *http://dx.doi.org/10.1787/5k3ttg88w8hh-en*.

City of Chicago (2008), *Chicago Climate Change Action Plan: Our City. Our Future*, City of Chicago, Chicago.

City of Copenhagen (2012), *Københavns Kommunes Skybrudsplan 2012*, City of Copenhagen, Copenhagen.

Climate-Adapt (n.d.), European Climate Adaptation Platform, Norway, Norwegian Climate Adaptation Programme, *climate-adapt.eea.europa.eu/countries/norway* (accessed 12 February 2015).

Council of the European Union (2013), "An EU strategy on adaptation to climate change: Council Conclusions", Communication from the Commission to the European Parliament, the Council, the European Economic and Social Committee and the Committee of the Regions, *http://register.consilium.europa.eu/doc/srv?l=EN&f=ST%2011151%202013%20INIT* (accessed 12 February 2015).

EEA (2014), *National adaptation policy processes in European countries: 2014*, European Environment Agency, Copenhagen.

Executive Office of the President (2013), *The President's Climate Action Plan*, The White House, Washington, DC.

French Government (2011), *National Plan: Climate Change Adaptation*, Ministry of Ecology, Sustainable Development, Transport and Housing, Paris.

Gagnon-Lebrun, F. and S. Agrawala (2006), "Progress on Adaptation to Climate Change in Developed Countries: An Analysis of Broad Trends", *OECD Papers*, Vol. 6/2, OECD Publishing, Paris, *http://dx.doi.org/10.1787/oecd_papers-v6-art8-en*.

Goodsite, M.E. et al. (2013), *White Paper: Climate Change Adaptation in the Nordic Countries*, Nordic Climate, Mitigation, Adaptation and Economic Policies Network (N-CMAEP), Norden Top-level Research Initiative, Oslo.

Government of Canada (2014), *Canada's Sixth National Report on Climate Change. Actions to meet commitments under the United Nations Framework Convention on Climate Change*, *http://unfccc.int/national_reports/annex_i_natcom/submitted_natcom/items/7742.php* (accessed 12 February 2015).

Hunter, P., Z. Burkitt and B. Trangmar (2010), "Local government adapting to climate change: Managing infrastructure, protecting resources, and supporting communities", in Nottage, R.A.C. et al. (eds.), *Climate change adaptation in New Zealand: Future scenarios and some sectoral perspectives*, New Zealand Climate Change Centre, Wellington.

ICLEI (Local Governments for Sustainability) and CEPS (The Centre for European Policy Studies) (2013), *Climate change adaptation: Empowerment of local and regional authorities, with a focus on their involvement in monitoring and policy design*, European Union, Committee of the Regions, Brussels.

IPCC (Intergovernmental Panel on Climate Change) (2014), *Climate Change 2014: Impacts, Adaptation, and Vulnerability*, Contribution of Working Group II to the Fifth Assessment Report of the Intergovernmental Panel on Climate Change, Cambridge University Press, Cambridge and New York.

IPCC (2007), *Climate Change 2007: Synthesis Report*, Contribution of Working Group I, II and III to the Fourth Assessment Report to the Intergovernmental Panel on Climate Change, Cambridge University Press, Cambridge and New York.

Klein, R.J.T. and S. Juhola (2013), "A Framework for Nordic Actor-Oriented Climate Adaptation Research", *NORD-STAR Working Paper 2013-01*, Nordic Centre of Excellence for Strategic Adaptation Research.

Lindley et al. (2011), *Climate Change, Justice and Vulnerability*, Joseph Rowntree Foundation, York.

Mackay, K. (2007), *How to Build M&E Systems to Support Better Government*, World Bank Independent Evaluation Group, Washington, DC.

Mullan, M. et al. (2013), "National Adaptation Planning: Lessons from OECD Countries", *OECD Environment Working Papers*, No. 54, OECD Publishing, Paris, *http://dx.doi.org/10.1787/5k483jpfpsq1-en*.

OECD (Organisation for Economic co-operation and Development) (2015), *National Climate Change Adaptation: Emerging Practices in Monitoring and Evaluation*, OECD Publishing, Paris, *http://dx.doi.org/10.1787/9789264229679-en*.

OECD (2009), *Integrating Climate Change Adaptation into Development Co-operation: Policy Guidance*, OECD Publishing, Paris, *http://dx.doi.org/10.1787/9789264054950-en*.

Pauw, P. and A. Pegels (2013). "Private Sector Engagement In Climate Change Adaptation in Least Developed Countries: An Exploration", *Climate and Development*, Vol. 5(4), *http://dx.doi.org/10.1080/17565529.2013.826130*.

Persson, Å. and R.J.T. Klein (2009), "Mainstreaming Adaptation to Climate Change into Official Development Assistance: Challenges to Foreign Policy Integration", in Harris, P.G. (ed.), *Climate Change and Foreign Policy: Case Studies from East to West*, Routledge, New York.

Pirard, P. et al. (2005), "Summary of the Mortality Impact Assessment of the 2003 Heat Wave in France", *Eurosurveillance*, Vol. 10(7).

Stalenberg, B. (2012), "Innovative Flood Defences in Highly Urbanized Water Cities", in Aerts, J. et al. (eds.), *Climate Adaptation and Flood Risk in Coastal Cities*, Earthscan Climate, New York.

Stalenberg, B. and Y. Kikumori (2008), "Urban flood control on the rivers of Tokyo metropolitan", in de Graaf, R. and F. Hooimeijer (eds.), *Urban Water in Japan*, Taylor & Francis Group, London.

Stenek, V., J.C. Amado and D. Greenall (2013), *Enabling Environment for Private Sector Adaptation: An Index Assessment Framework*, International Finance Corporation, Washington, DC.

UNFCCC (n.d.), Nairobi Work Programme Private Sector Initiative – database of actions on adaptation, United Nations Framework Convention on Climate Change, *http://unfccc.int/adaptation/workstreams/nairobi_work_programme/items/6547.php* (accessed 12 February 2015).

US Department of State (2014), *United States Climate Action Report 2014 (2014 CAR)*, US Department of State, Washington, DC.

Wolstenholme, R. (2010), *The role of parliamentarians in strengthening the climate change agenda: Scotland report*, International Institute for Environment and Development, London.

Climate Change Risks and Adaptation
Linking Policy and Economics

Chapter 3

Overview of costs and benefits of adaptation at the national and regional scale

This chapter reviews the latest evidence on the costs and benefits of adaptation, and draws out some of the key findings and emerging insights. It explores the use of information on the costs and benefits of adaptation to justify the case for action and prioritise resources to deliver the greatest benefits. Results of national and global studies are provided. The latest estimates are provided for the following sectors and risks: sea-level rise, coastal flooding and storms; river, surface water and urban flooding; water supply and management; infrastructure; agriculture; health; biodiversity and ecosystem services; business, services and industry.

Key messages

- The information base on the costs and benefits of adaptation has significantly evolved in recent years. It has moved beyond the previous focus on coastal areas to include water management, floods, agriculture and the built environment. However, gaps remain for ecosystems and business, services and industry.
- The methods for identifying options and assessing costs and benefits are also changing. There is an increasing use of new approaches that aim to support decision making under uncertainty, and a focus on early low-regret options. This leads to a different suite of options, including a focus on capacity building and non-technical options, and differences in the timing and phasing of options.
- Improved information is also available on the aggregate costs of adaptation. Recent implementation and policy orientated studies indicate higher costs than the previous review, because of existing policy objectives and standards, the need to consider multiple risks and uncertainty, and additional opportunity and transaction costs associated with policy implementation.
- While important gaps exist in the empirical evidence, and there are issues of transferability and the limits of adaptation, the new evidence base provides an increased opportunity for sharing information and good practice.

The costs and benefits of adaptation

The analysis of the costs and benefits of adaptation has an important role in justifying the case for action, and for prioritising available resources to deliver greater social, environmental and economic benefits. This information is relevant at the global level, as an input to the discussion on international financing needs. It is also relevant for national adaptation plans to allow efficient, effective and equitable strategies, and for local and project level adaptation, as a key input to appraisal.

In theory, a common framework can be used for the analysis of costs and benefits at all three geographic levels (Boyd et al., 2004; Stern et al., 2006), and this has been widely adopted. This framework first assesses the impacts and economic costs of climate change, including slow onset trends and changes in extreme events. It then assesses the potential costs and benefits of adaptation to reduce these impacts. This information can be used to assess the economic effectiveness of adaptation, i.e. whether the economic benefits of adaptation outweigh the costs. It can also be used to compare alternative adaptation options. There is, however, an additional step to undertake in this analysis. This assesses the residual impacts of climate change after adaptation, noting that it will rarely be completely effective – or even technically possible – to remove impacts completely. The most effective (or even economically optimal) level of adaptation will therefore be a balance between the costs of adaptation measures, the benefits of adaptation measures and residual impacts.

More recently, the discussion around climate change has shifted towards a focus on risk (IPCC, 2014a), which is reflected in this report. This leads to some changes in the terminology compared to the framework above, with future climate risks (rather than impacts) and residual risks remaining after adaptation. In a risk framework, the costs of adaptation referred to below are equivalent to investments in risk reduction. More fundamentally, it has led to a change in the framing around adaptation, moving away from the previous impact-assessment framework towards iterative climate risk management.

Against this background, this chapter outlines some of the main issues with assessing the costs and benefits of adaptation, including an overview of the current state of the evidence base, set in the context of this new risk framework. It draws on the research, analysis and review of the ECONADAPT project, funded by the European Union's Seventh Framework Programme for research, technological development and demonstration[1] and co-funding provided by the UK Department for International Development and Canada's International Development Research Centre.[2] The chapter briefly introduces the challenges involved in estimating the costs and benefits of adaptation. It then assesses different evidence lines at global, national and sectoral levels. Significantly, this moves beyond the previous framing of adaptation, to consider early, practical adaptation under uncertainty. Finally, the findings from the review are highlighted and gaps identified. A more detailed review – including more detailed estimates for developing counties and more information and analysis of the studies and cost estimates – is provided in a supporting ECONADAPT report (2015) on the *Costs and Benefits of Adaptation.*

From theory to practice

A number of methods have been developed to derive estimates of the costs and benefits of adaptation (Watkiss and Hunt, 2010), though these have primarily used the impact-assessment framework described above. While the approach is straightforward, there are number of challenges in putting it into practice (Füssel and Klein, 2006; UNFCCC, 2009; UNEP, 2014; ECONADAPT, 2015).

- **Estimating future climate risks and adaptation benefits:** It is difficult to estimate the future impacts and economic costs of climate change, due to the wide range of potential risks, the scientific and economic information available, data gaps and modelling constraints. These issues are amplified when considering adaptation costs and benefits, especially given the large number of potential adaptation options that exist.
- **Uncertainty:** The challenges above are made more difficult because of the large uncertainty associated with future climate change. At the current time it is not clear what future emission pathway the world is on, and even if this were known – significant climate model uncertainty would remain. Taking account of this uncertainty has two consequences: it makes it harder to estimate the scale of the impacts of climate change and the benefits of adaptation; and it increases adaptation costs relative to a situation where people are assumed to be able to predict the future.
- **Framing:** The costs and benefits of adaptation are determined by the framework that is used and the objectives that are set (e.g. whether the optimal level is based on economic efficiency versus a defined level of acceptable risk). These vary with context, country and across stakeholder groups. This means it is very difficult to provide a definitive cost of adaptation. There are also additional issues around the distributional effects (and

equity) of climate change over time and between groups, and whether these are accounted for in analysis of impacts and adaptation.

- **Baselines and timescale:** The baseline assumptions, and the future timescales under investigation, lead to large variations in estimates. The choice of discount rate is of particular relevance in this context, as it affects the weight put on benefits occurring in the future. There is the further issue of the existing adaptation deficit (the gap between the current state of a system and a state that minimises adverse impacts from existing climate variability), as adaptation to future climate change will be less effective if this deficit has not first been addressed (Burton, 2004). This is a particular problem for developing countries, but even OECD countries have adaptation deficits or are close to the limits of coping with current climate variability (ASC, 2011).
- **Scale and boundaries**: The impacts of climate change in one area will spillover into other areas through mechanisms such as trade and financial flows. These can only be modelled at a global scale, but can affect costs reported at the regional, national or local scale.

The assumptions used to address these challenges can have a large impact on results. A consequence of this is that the results of any studies – and the estimates of the costs and benefits of adaptation they produce – have the potential to be misleading if viewed in isolation. It is important for any study to be transparent about the assumptions used and implications of these on potential decisions.

Finally, one of the key aims of investigating the costs and benefits of adaptation is to help allocate resources, to inform national adaptation planning by governments through to local decisions. The impact assessment (I-A) framework outlined above reflects a stylised model of reality: It calculates technical costs, which are used to estimate the reduction in future damages. While such studies are useful for raising awareness, and generating headline estimates of the costs and benefits, they are less useful for practical (early) adaptation as they are highly theoretical. More recent studies highlight an emerging set of challenges in addition to those listed above (ECONADAPT, 2015):

- **Adaptive capacity:** Recent studies have highlighted the need to build adaptive capacity and focus on the process of adaptation, as well as delivering adaptation options (Downing, 2012). Building adaptive capacity involves sharing information, research, monitoring, raising awareness, education and training, and other institutional and organisational activities (UKCIP, 2006). It is a key priority for government in creating the enabling environment for adaptation. However, it is often omitted in technical studies, and its indirect nature makes it difficult to assess costs and benefits. There has been some progress in considering the value of information in relation to climate services and adaptation (Clements, 2013; Macauley, 2010) but this remains a priority.
- **Wider issues and policy context:** As adaptation moves towards implementation, there is a greater need to include wider (non-climatic) drivers and existing policy in analysis. Earlier studies, particularly those that use an I-A framework, ignore these factors (Füssel and Klein, 2006; UNFCCC, 2009), yet these are often more influential than climate change, particularly in the short-term.
- **Autonomous and private sector adaptation:** Autonomous adaptation will arise for many of the risks of climate change, but this has rarely been quantified, as there is a lack of empirical evidence, and it is difficult to include this in most impact and modelling assessments (with exceptions for agriculture and energy demand). While this is likely to

be more reactive than planned, it remains a priority, especially for considering the potential actions of the private sector and how this can affect the nature and extent of government intervention.

- **Opportunity and transaction costs associated with policy implementation:** Impact assessment estimates are usually based on technical costs, but there can be important opportunity costs associated with measures, as well as the transaction costs to introduce and implement the required measures. These can have a major influence on the choice of options and the aggregate estimates of adaptation costs and benefits.
- **Cross-sectoral, cross-cutting and macroeconomic interactions:** Cross-sectoral and cross-cutting effects of adaptation measures – and likewise ancillary costs and benefits – are rarely taken into consideration in adaptation costing, but this is becoming increasingly important in moving to implementation. It is also clear that including such effects can significantly affect the ranking of adaptation measures (Skourtos, Kontogianni and Tourkolias, 2013). At the macroeconomic level, the effects of climate impacts in one sector will feed through across the economy, though this is often omitted in studies. An emerging priority is to understand these wider economic costs of adaptation and their importance for public finances, GDP, employment, investment and so on.
- **Decision-relevant timescales:** The impacts from climate change will become most apparent from the middle of the century onwards, and this is when many studies focus their efforts. However, there is less policy relevance in estimating the future costs of adaptation in 2040 and beyond. Instead, the key issue is what to do in the next decade or two, both to address early changes and to prepare for the longer-term.
- **Transformative adaptation and the limits of adaptation:** Incremental adaptation helps to maintain the essence and integrity of a system or process at a given scale, while transformational adaptation changes the fundamental attributes of the system (IPCC, 2014). There is, as yet, little economic evidence on transformational adaptation. This is increasingly important given the recognition of the limits of adaptation (Adger et al., 2007), including physical and ecological limits, technological limits, financial barriers, information and cognitive barriers, and social and cultural barriers. These limits are omitted in most current studies and are a critical gap.

In response to the various challenges of adaptation costing, and these emerging issues, the framing of adaptation has changed in recent years, moving away from a focus on science-first and impact assessment. In particular, the use of iterative climate risk management to consider uncertainty has emerged (see Chapters 4 and 6), which considers climate and non-climate risks as a dynamic set of risks, and identifies phased adaptation responses. These changes have important implications for the costs and benefits of adaptation. It is now difficult to compile and compare estimates, because of the different approaches being used. Studies use different methods, objectives, metrics and assumptions, and often focus on different time periods, and are conducted at different scale and geographical resolution. No method is absolutely right or wrong and they all have strengths and weaknesses according to the objectives of the exercise and the specific application. However, there is a major difference between earlier and later studies, and they are reported separately in the review below. The focus is therefore on assessing the state-of-the-art and key lessons, rather than providing absolute estimates of the costs of adaptation.

Current state of the literature

Over the past few years, there have been several reviews of the costs and benefits of adaptation (EEA, 2007; OECD, 2008; UNFCCC, 2009; Agrawala et al., 2011a; Markandya et al., 2014; Chambwera et al., 2014). These report that the evidence base is relatively small and stress that deriving estimates involves many challenges, and that different studies have used different methods, time scales, climate scenarios, objectives and assumptions, making comparisons difficult. As estimates of the costs (and benefits) of adaptation are conditional on the assumptions and data sources used, there are large variations in estimates for countries and sectors between studies. Estimates will also be highly context specific and will vary with existing national, sectoral and individual preferences, as well as analysis and choice of metrics.

There are, however, a growing number of national level initiatives, varying from one or two key sectors through to economy wide assessments. There are also more sectoral studies that focus on early adaptation. These two factors have led to a much larger number of studies – and evidence – on the costs and benefits of adaptation. A recent major review and compilation of these studies (ECONADAPT, 2015) has identified around 500 studies and an early review of these form the basis of this chapter. While the review has aimed to be comprehensive as possible, there will inevitably be additional relevant studies, especially given the rapid emergence of this literature.

The review as evidence is summarised in Table 3.1 below, taken from the supporting ECONADAPT report. While coastal risks remain the most comprehensively covered, more literature has emerged for other risks and sectors. There is also a greater geographical coverage, as shown in Figure 3.1. However, while the evidence base is growing, and the coverage has expanded, it remains partial for all sectors (regarding, for example, impacts,

Table 3.1. **Updated quality of the coverage of the sectors in the adaptation literature**

Risk/Sector	Coverage/Discussion	Cost estimates	Benefit estimates
Coastal zones and coastal storms	Comprehensive coverage (flooding and erosion) at global, national and local level in impact assessment studies. Good evidence base on early low regret options and iterative adaptive management including policy studies and decision making under uncertainty (real options).	✓✓✓	✓✓✓
Floods including infrastructure	Growing number of adaptation cost and benefit estimates (impact assessment studies) in a number of countries and local areas, particularly on river flooding. Evidence base emerging on low regret options and non-technical options. Some applications of decision making under uncertainty.	✓✓	✓✓
Water sector management including cross-sectoral water demand	A recent focus on supply-demand studies at the national level, but a range of global, river basin or local studies available. Focus on supply, engineering measures; less attention to demand, soft, and ecosystem-based measures. Some examples of decision making under uncertainty, particularly robust decision making, with policy relevant studies.	✓✓	✓
Other infrastructure	Several studies on road and rail infrastructure. Examples of wind storm and permafrost.	✓	✓
Agriculture (multi-functionality)	High coverage of the benefits of farm level adaptation (crop models), and some benefits and costs from impact assessment studies at global and national level. Evidence base emerging on potential low regret adaptation, including climate smart agriculture options (soil and water management)	✓✓	✓✓
Over-heating (built environment, energy and health)	Good cost information on heat-alert schemes and some cost-benefit studies for future climate change. Increasing coverage of autonomous costs[1] associated with cooling from impact assessment studies (global and national). Growing evidence base on low-regret options for built environment (e.g. passive cooling).	✓✓	✓

Table 3.1. **Updated quality of the coverage of the sectors in the adaptation literature** (*cont.*)

Risk/Sector	Coverage/Discussion	Cost estimates	Benefit estimates
Other health risks	Increasing studies of preventative costs for future disease burden (e.g. water, food and vector borne disease), but coverage remains partial.	✓	✓
Biodiversity/ecosystem services	Low evidence base, with a limited number of studies on restoration costs and costs for management of protected areas for terrestrial ecosystems.	✓	
Business, services and industry	Very few quantitative studies available, except for tourism, some focusing on winter tourism and some on autonomous adaptation from changing summer tourism flows.[1]	✓	

Note: See main text for discussion and caveats.
1. can be considered an impact or as autonomous adaptation. (i.e. unplanned).
Key:
✓✓✓ Comprehensive coverage at different geographical scales and analysis of uncertainty.
✓✓ Medium coverage, with a selection of national or sectoral case studies.
✓ Low coverage with a small number of selected case studies or sectoral studies.
The absence of a check indicates extremely limited or no coverage.
Source: ECONADAPT (2015), "The Costs and Benefits of Adaptation", results from the ECONADAPT Project, ECONADAPT consortium.

Figure 3.1. **National level adaptation cost studies**

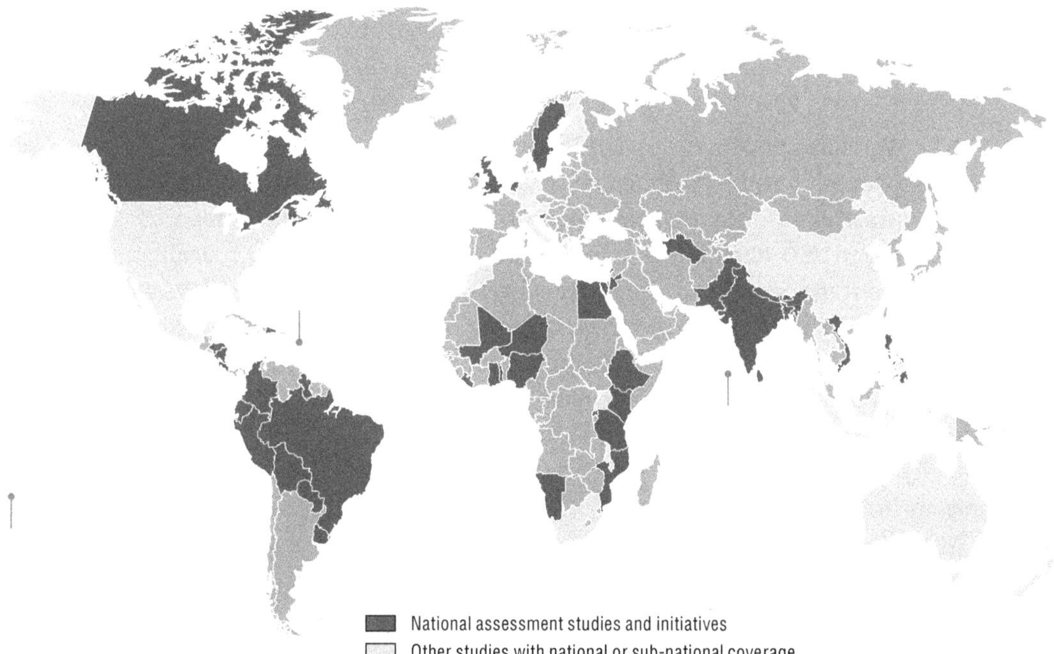

Source: ECONADAPT (2015), "The Costs and Benefits of Adaptation", results from the ECONADAPT Project, ECONADAPT consortium.

spatial and temporal scales, benefit estimates) and major gaps exist for ecosystems, business and industry. These studies also use a diverse set of methods, socio-economic assumptions, cost metrics and benefit categories, as well as discount rates. This makes inter-comparison difficult. Finally, the number of more policy orientated and iterative studies remains low. The review therefore cautions against the simple reporting of the costs of adaptation and further analysis of transferability is a key research priority. The following sections compile the evidence and lessons, starting with the global assessments, and then providing national and risk- and sector-based estimates.

Global estimates

Early estimates of the global costs of adaptation focused on the period to 2030. Six assessments were undertaken (OECD, 2008), which primarily estimated costs using an investment and financial flow analysis method. This applies an adaptation cost "mark-up" to future investment plans to take account of future climate change. These studies have the advantage of grounding the analysis in current policy and plans, but they have a less direct link to future climate change and uncertainty. The most comprehensive was the study by the United Nations Framework Convention on Climate Change, which estimated global adaptation investment needs at USD 50 to USD 170 billion per year by 2030 (equivalent to 0.06-0.2% of projected GDP), primarily associated with infrastructure protection in developed countries (UNFCCC, 2007). This study estimated that only USD 30 to USD 70 billion per year would be required for adaptation investments in developing countries. However, a critique by Parry et al. (2009) argued that this study underestimated adaptation costs by a factor of 2 to 3, and stressed that many sectors and impacts were not included.

A subsequent assessment – focusing on developing countries – was undertaken by the World Bank in the Economics of Adaptation to Climate Change (EACC) study (2010). This study used impact assessment to estimate the economic costs of climate change, then estimated the costs of adaptation to achieve pre-climate levels of welfare. The total adaptation cost for developing countries was estimated at USD 70 billion to USD 100 billion per year (using 2005 nominal values). This estimate reflects the average projected costs between 2010 and 2050 for a 2 °C warmer world.

The World Bank estimate is slightly higher than the UNFCCC study for similar regions. This study found that the projected costs were highest in East Asia and the Pacific Region, and for infrastructure, coastal zones and the water sector. The study reported costs rising from USD 60 to USD 70 billion per year by 2010-19 to USD 90 to USD 100 billion per year by 2040-49. The study considered two climate futures: i) minimum and maximum temperature, and ii) "wetter" and "drier" projections. It found higher costs with wetter scenarios due to infrastructure impacts. The choice of aggregation rule also affected the size of the estimates, notably whether gains from climate change were added to adaptation costs. However, as the report acknowledges, adaptation costs were still calculated as though decision makers know the future with certainty. Moreover, many of the criticisms of Parry et al. (2009) still apply, including that the coverage of impacts and sectors is partial.

Finally, an alternative set of global estimates has been derived from global economic integrated assessment models (IAMs). These quantify the economic impacts of climate change, and in some cases, the costs and benefits of adaptation, see Box 3.1. They tend to focus on the long-term, and have been used to assess mitigation and adaptation costs. Most adaptation estimates are based on the AD-RICE or PAGE models (see Parry et al., 2009; de Bruin et al., 2009; Agrawala et al., 2011a; Bosello et al., 2013; Dellink et al., 2014). More recent assessments, notably de Bruin (2014), assess how adaptation costs could vary along different emission pathways, finding costs could be around twice as high in a 4 °C scenario than a 2 °C one, even by 2050. These IAMs have also been applied at the continental level, including by the Asian Development Bank (ADB, 2014) for the Economics of Climate Change for South Asia.

Box 3.1. Consequences of climate change damages for economic growth: OECD assessment

The OECD's dynamic global general equilibrium model, ENV-Linkages, assesses the consequences of a selected number of climate change impacts in the various world regions at the macroeconomic and sectoral level. The analysis estimates that the climate change impacts on annual global GDP are projected to increase over time, leading to a global GDP loss of 1.0% to 3.2% by 2060 for the most likely equilibrium climate sensitivity range, with the greatest impacts on the agricultural sector. Nevertheless, some impacts and risks from climate change have not been quantified in this study, such as large-scale disruptions. They will potentially have large economic consequences, and on balance the costs of inaction estimated in this study are likely to underestimate the full costs of climate change impacts.

Source: Dellink, R. et al. (2014), "Consequences of Climate Change Damages for Economic Growth: A Dynamic Quantitative Assessment", *OECD Economics Department Working Papers*, No. 1135, OECD Publishing, Paris, *http://dx.doi.org/10.1787/5jz2bxb8kmf3-en*. OECD (2015), "The economic consequences of climate change", forthcoming September 2015.

National estimates

A large number of national studies that consider the costs and benefits of adaptation have emerged in recent years. An indicative mapping of these studies is shown in the Figure 3.1, compiled by the ECONADAPT project (2015) and summarised below.

OECD countries

A number of national level assessments have considered adaptation costs and benefits in OECD countries. In the survey undertaken for this report, Netherlands, the United Kingdom, and Slovenia reported that economic assessments were included in their national adaptation programmes. The first two of these reflect some of most advanced examples globally. They have evolved over many years, from early impact assessment, to analysis of adaptation options and possible costs (UKCIP, 2006; van Ierland et al., 2006), and finally to advanced iterative frameworks with the Delta programme in the Netherlands (Delta Programme, 2011; Delta Programme, 2014; Eijgenraam et al., 2014) and the United Kingdom's Economics of Climate Resilience and the National Adaptation Programme (Watkiss and Hunt, 2011; Frontier Economics, 2013; HMG, 2013).

There are also studies in other European countries that have costed adaptation. The analysis in Sweden (SCCV, 2007) presented investment and financial flow costs for several sectors, and the Bank of Greece study (BoG, 2011) assessed costs for an adaptation scenario. Earlier work in Italy looked at the economics of adaptation and some options (Carraro and Sgobbi, 2008) and a study in Germany undertook cost-benefit analysis on 28 potential adaptation options (Tröltzsch et al., 2012). At the European level, there are academic studies that have considered several sectors, such as the PESETA (Ciscar et al., 2011) and ClimateCost (Watkiss et al., 2012) studies, as well as sector specific estimates (see later discussion).

In the Americas, some of the earliest work on the costs and benefits of adaptation was in Canada (Dore and Burton, 2001). This was followed by numerous studies in specific sectors and regions (see Environment Canada, 2006; NRC, 2007; NRTEE, 2011). Similarly, in the United States, there are national level studies that provide estimates in specific sectors or regions. While the recent US National Assessment (2014) did not compile national adaptation costs, a recent review (Sussman et al., 2014) summarised the current state of knowledge. There are also many state-level climate change-specific adaptation actions that focus on planning and include an analysis of adaptation costs.

While there is not sufficient information to assess the total costs of adaptation in OECD countries, country level information is emerging. As examples, the annual costs for future flood protection and flood-risk management in the Netherlands have been estimated to be in excess of EUR 1 billion per year (Delta Commissie, 2008) and similar annual costs have been estimated for the United Kingdom (Foresight, 2004; EA, 2008; EA, 2011; ASC, 2014). In the United States, estimates suggest that adaptation costs could be as high as tens or hundreds of billions of dollars per year by the middle of this century (Sussman et al., 2014). Finally, the picture continues to evolve and a growing number of countries are starting to consider the costs and benefits. For example, recent adaptation cost studies or initiatives have commenced in Austria, Spain and Mexico.

Developing countries and emerging economies

In recent years, a number of initiatives have emerged that estimate the costs of adaptation in developing countries and emerging economies. While the focus of this chapter is on OECD countries, these studies provide a large additional source of evidence. They also provide important practical lessons from the early implementation of adaptation, which is advancing rapidly in many of these countries. The estimates are also highly relevant to OECD countries because they can help to inform international development assistance for adaptation. The evidence base includes:

- The World Bank's EACC country studies (in Bangladesh, Bolivia, Ethiopia, Ghana, Mozambique, Samoa, and Viet Nam), which used impact-assessment, but also provided more detailed (bottom-up) assessment and considered economy wide effects (World Bank, 2010).
- The United Nations Development Programme (UNDP) Investment and Financial Flows initiative (UNDP, 2011) estimated the costs of adaptation through to 2030 in 15 countries (Bangladesh, Colombia, Costa Rica, Dominican Republic, Ecuador, Gambia, Honduras, Liberia, Namibia, Niger, Paraguay, Peru, Togo, Turkmenistan, and Uruguay) for 1 or 2 key sectors.
- The UNFCCC National Economic, Environment and Development Study (NEEDS) assessed the short- and long-term costs of adaptation and financing needs in Egypt, Ghana, Jordan, Lebanon, Maldives, Mali, Philippines and Nigeria (UNFCCC, 2010).
- Additionally, a large a number of other regional and country level studies have been undertaken, including in Bangladesh, Brazil, Bhutan, China, Ethiopia, Guyana, Kenya, India, Indonesia, Maldives, Nepal, Philippines, Peru, Rwanda, Samoa, South Africa, Sri Lanka, Tanzania, Thailand, Uganda and Viet Nam, as well as Caribbean and Central America regions. Details are provided in ECONADAPT (2015).

The evidence from these studies provides some new insights. They suggest that adaptation costs for these countries are potentially higher than reported in the EACC study in the period up to 2030 and beyond (UNEP, 2014). This can, in part at least, be explained by the coverage of the risks, the consideration of higher levels of temperature change (beyond 2 °C), the challenge of uncertainty, and the consideration of the existing adaptation deficit. It also reflects the emerging experience that implementation of measures entails costs beyond the technical cost of the measure itself.

Risk- and sector-based estimates

The previous OECD (2008) review reported estimates on a sector by sector basis. This provides a useful entry point, but in many cases adaptation is a response to a cross-sectoral risk (e.g. floods or heat waves), thus there is the potential for co-ordinated responses to share costs, generate co-benefits and address potential conflicts. For this reason, it is useful to consider cross-sectoral risks as well as individual sectors, and this is the format used below. As highlighted above, this section summarises the state of evidence and potential lessons and insights. More detailed cost estimates are presented in ECONADAPT (2015).

Sea level rise, coastal flooding and storms

There are a number of risks from climate change on coastal zones, including sea level rise, storm surges and wind-storms, flooding, loss of land, coastal erosion, salt water intrusion and impacts on coastal wetlands. In response, there is a broad set of adaptation options – generally based around protection, retreat or accommodation (IPCC, 2007). There is a comprehensive evidence base on these costs and benefits (OECD, 2008), and this has increased further in recent years.

A large number of studies have used sector impact assessment models to assess adaptation costs and benefits, many using the Dynamic Interactive Vulnerability Assessment (DIVA) model, which assesses the costs of physical barriers (dikes) to address flood risks and shoreline management (beach nourishment) to address coastal erosion (Hinkel and Klein, 2009). DIVA has been used for global studies (UNFCCC, 2007; World Bank, 2011), regional studies (e.g. in Africa [Brown et al., 2009] and Europe [Hinkel et al., 2010]) and national studies (e.g. for individual European countries [Brown et al., 2011] and in Mozambique [World Bank, 2010b]; Ghana [World Bank, 2010c]; Kenya [SEI, 2009]; Peru [IBD/ECLAC, 2014]; Tanzania [GCAP, 2011]; and India [Markandya and Mishra, 2010]). The most recent runs of the model suggests that estimated global annual investment and maintenance costs of protecting the coast to 2100 are in the range of USD 12-31 billion to USD 27-71 billion for low and high warming scenarios (Hinkel et al., 2014). The additional adaptation costs associated with coastal erosion (beach and shore nourishment) are estimated at a further USD 1.4-5.3 billion per year across low, mid and high scenarios (Hinkel et al., 2013).

Similar types of impact assessment studies have been undertaken in a number of countries, including Canada (Stanton, Davis and Fencl, 2010), Brazil (Margulis and Dubeux, 2010), the United Kingdom (Evans et al., 2004), the United States (Neumann et al., 2011), Germany (Tröltzsch et al., 2012); and there are also similar studies at local level. All of these studies find that coastal adaptation reduces damages significantly at relatively low cost and leaves low residual damages: it thus has high benefit to cost ratios, which generally increase throughout the 21st century. Annual adaptation costs are also generally a low proportion of GDP: For example, Agrawala et al. (2011b) estimated that it accounts for less than 0.1% of GDP though this does vary with country and region.

The cost estimates, however, tend to exclude adaptation costs linked to wind storm damage, salinization, port infrastructure, tourism, coastal and marine ecosystems, and often dike maintenance costs. These estimates also often assume good levels of existing protection and no adaptation deficit (Brown et al., 2011) – the latter a key issue for developing countries. There are some studies that address these gaps. The World Bank's EACC study (2010) estimated that costs were modest (USD 0.2-0.5 billion) when compared to the costs of sea defences. Local studies also emerging (e.g. the IFC (2011) report on the port of Cartagena,

Colombia) that address some of these issues. Similarly, studies have assessed the adaptation costs for tropical windstorm damage under climate change, including in Florida (e.g. RMS, 2009), the Caribbean (ECA, 2009; CCRIF, 2010) and Samoa (World Bank, 2009c). Nonetheless, the coverage of adaptation costs is still partial.

There are also coastal studies that are more policy focused. These reveal some new insights when compared to the impact assessment (I-A) studies described above. First, estimates of coastal adaptation costs and benefits vary with the level of protection (the objective). Earlier studies assume modest protection levels that are below existing standards in some OECD countries (e.g. the Netherlands), and in some cities (e.g. in London). In such cases, maintaining current protection levels will lead to higher adaptation costs. Second, I-A studies assume foresight – the models are run for one scenario at a time – and thus do not consider uncertainty. Related to this, studies that consider more extreme sea level rise (i.e. projections of 1 metre or more) report sharp increases in damage and adaptation costs (e.g. globally [Vafeidis et al., 2011] and in Europe [Brown et al., 2011]). Third, major cities may face much higher adaptation costs, especially for port-river cities which require highly engineered protection. As an example, the costs of protecting London against future sea level rise may require the construction of an additional flood barrier later this century (under a high sea-level rise scenario), which could cost GBP 6-7 billion (EA, 2009; EA, 2011). Hallegatte et al. (2013) analysed 136 global coastal cities and reported indicative adaptation costs of USD 350 million per year per city, or approximately USD 50 billion per year in total. Finally, I-A studies assume highly effective adaptation and ignore the costs of developing and implementing policies. Emerging policy studies in the Netherlands and United Kingdom (discussed above) indicate national adaptation costs that are many times higher than the impact assessment studies for the same countries (e.g. compared to Brown et al., 2011) for the various reasons above, i.e. higher risk protection levels, the consideration of uncertainty, more complex adaptation responses and policy costs.

While these earlier impact assessment studies provide critical information and context, recent studies have moved towards the analysis of early low-regret (see Box 3.2) and iterative adaptation. A number of these draw on existing disaster risk management and soft or non-technical options. Promising early low-regret options include (Mechler, 2012; ECONADAPT, 2015):

Box 3.2. **What are low- and no-regret adaptation options?**

Numerous studies recommend that no- and low-regret actions are a good starting point for early adaptation, as they offer benefits now and lay the foundation for future resilience (UKCIP, 2006; Watkiss and Hunt, 2011; Ranger and Garbett-Shiels, 2012; IPCC SREX, 2012). No-regret adaptation is defined (in the IPCC glossary) as adaptation policies, plans or options that "generate net social and/or economic benefits irrespective of whether or not anthropogenic climate change occurs". This includes options that address the current adaptation deficit (e.g. disaster risk management), options that are more efficient and generate cost savings (e.g. improving irrigation efficiency) or options that address existing problems (e.g. reducing post-harvest losses), though many of these are similar to development. There is, however, no agreed definition of low-regret options, and definitions include: i) options that are no-regret in nature, but have opportunity, transaction or policy costs; ii) options that have benefits (or co-benefits) that are difficult to monetise; iii) low cost measures that can provide high benefits if future climate change emerges; iv) options

Box 3.2. **What are low- and no-regret adaptation options?** *(cont.)*

that are robust or flexible, and thus help with future uncertainty. DFID (2014) – and this report – use a pragmatic definition of "low-regret" – that focuses on promising options for early adaptation. This includes options that are effective in addressing the current adaptation deficit, but also future-orientated, low-cost options that build resilience, flexibility or robustness, as well as capacity building and the benefits it provides through the value of information.

Source: UKCIP (UK Climate Impacts Programme) (2006), Identifying adaptation options, UKCIP, London; Watkiss, P. and A. Hunt (2011); Method for the UK Adaptation Economic Assessment (Economics of Climate Resilience), Final Report to Defra, May 2011. Deliverable 2.2.1; Ranger. N. and S.-L. Garbett-Shiels (2011), How can decision-makers in developing countries incorporate uncertainty about future climate risks into existing planning and policy-making processes?, Policy Paper, Grantham Research Institute on Climate Change and the Environment, London.; IPCC (2012), Managing the Risks of Extreme Events and Disasters to Advance Climate Change Adaptation, A Special Report of Working Groups I and II of the Intergovernmental Panel on Climate Change, Cambridge University Press, Cambridge and New York; DFID (2014), Early Value-for-Money Adaptation: Delivering VfM Adaptation using Iterative Frameworks and Low-Regret Options, DFID, London.

- **Climate services, forecasting and early warning systems:** These have high benefit to cost ratios (World Bank, 2011), as shown by example in the United States on hurricane risk (Lazo, Rice and Hagenstad, 2010; Lazo and Waldman, 2011; Considine et al., 2004) and in developing countries (e.g. Bangladesh [Paul, 2009] and South-East Asia [Subbiah, Bildan and Narasimhan, 2008]). Benefits generally increase under climate change (ECA, 2009) though this depends on the risk.
- **Disaster risk management and emergency/contingency plans:** This includes forums/ institutional strengthening, awareness raising, response plans and emergency infrastructure including shelters and rescue centres. These measures have high benefit to cost ratios for current risks and future climate change (e.g. Cartwright et al., 2013 in Durban).
- **Natural coastal buffer zones:** These include mangrove conservation, replanting and restoration, and similar measures for seagrass and coral, as well as shoreline restoration and marine protection sites. Such ecosystem based measures have been prioritised (with high benefit to cost ratios) in many studies, (e.g. in Samoa [ECA, 2009] and the Caribbean [CCRIF, 2010]). However, in high income countries, the costs of mangrove restoration can be very high (e.g. in the United States [World Bank, 2011]) and there are potential opportunity (land) or policy (enforcement) costs.
- **Risk transfer including insurance, reserve funds and risk pools/facilities:** These include a variety of mechanisms for risk transfer, and are particularly important for low-probability, high-consequence events (IPCC, 2012; CCRIF, 2010; Mechler, 2012).

The list of low-regret options above have upfront benefits – and provide enhanced resilience for the future – though on their own, they may not be sufficient to address more extreme risks from longer-term change (see World Bank, 2011). There are also a set of responses that build early resilience to longer-term change, though these tend to involve more site specificity and thus their low-regret characteristic varies. Examples include:

- **Climate risk screening:** When applied to major infrastructure developments, climate risk screening can be used to consider location and design and has been found to have high cost-effectiveness if included at the design stage, because of the avoided reconstruction costs from floods and storms (e.g. ADB [2005] in Micronesia and Cook

Islands). The siting of critical infrastructure such as hospitals and water treatment facilities away from high risk areas is also a low regret option. In some cases, a degree of over-design to higher protection levels is justified, because of their importance following a disaster (World Bank, 2011).

- **Land-use planning and set-back zones:** Some studies report that coastal zoning or back away areas or lines (where development is prohibited) have high benefit to cost ratios for hurricane protection under current climate variability and future climate change (e.g. in the Caribbean [CCRIF, 2010] and in Samoa [ECA, 2009]) and storm-surge (Cartwright et al., 2013 in Durban). However, in middle income and OECD countries, these involve high opportunity costs of land.
- **Building codes:** While building codes are often cited as a potential low-regret option (e.g. IPCC, 2012), the picture varies. Some studies find high benefits (e.g. in Florida [ECA, 2009] and in Samoa [World Bank, 2010]), but others report low benefit to cost ratios (e.g. in the Caribbean [CCRIF, 2010; Hochrainer-Stigler et al., 2011]) due to differences in risk levels, the costs of resilience, existing cost and asset life-time, and assumed discount rates.

In OECD countries, there is increasing interest in alternatives to engineered coastal defences. There are studies that assess the costs and benefits of spatial planning options (see de Bruin et al., 2014). There is also an increasing focus on soft or ecosystem-based (green) protection (e.g. sand dunes, offshore sand banks, and sand engines, as well as managed retreat and coastal wetlands). These approaches have potential advantages, providing co-benefits and flexibility against future uncertainty. A number of studies have assessed their costs and benefits. A cost-benefit assessment of salt marshes in the Netherlands found that ecosystem-based approaches were less expensive than traditional options over the longer term (net present value) and in terms of construction costs, but that they were more expensive in terms of management and maintenance costs alone (De Bel, Schomaker and van Herpen, 2011). De Bruin et al. (2012) looked at future sea level rise in the Netherlands and compared sand dunes against hard structural protection: Sand dunes offered greater flexibility and lower capital costs, but maintenance costs were higher. The choice of discount rate is therefore critical in choosing between these options. There are also recent studies that have assessed flood management strategies under climate change, comparing low-regret and green alternatives against large-scale flood protection infrastructure. Examples include the analysis of flood management New York in the United States (Aerts et al., 2013; Aerts et al., 2015), which compared wetland restoration and increased building codes and recommended a hybrid solution, combining protection of critical infrastructure and resilience measures that could be upgraded over time, at least in the medium term.

Finally, a number of OECD countries are adopting adaptive management, looking at the overall adaptation pathway, from short- to long-term responses. In the Netherlands the Delta programme has advanced short-term measures that increase adaptability (flexibility) and resistance to extreme events (robustness), and delay tipping points. Most recently, the programme has moved to dynamic adaptation pathways (Delta Programme, 2014) and dynamic cost-benefit analysis (Kind, 2014; Eijgenraam et al., 2014). Similarly, the Thames Estuary 2100 project used an iterative approach to consider future protection for London, considering a portfolio of options linked to enhanced monitoring (EA, 2009; EA, 2011). There are also a number of studies that have applied decision making under uncertainty tools (see Chapter 6), including the use of real options analysis for dike heightening in the

Netherlands (van der Pol et al., 2013), an analysis of hard infrastructure, restoration of mangroves and coastal zone management options in Mexico (Scandizzo, 2011) and the value of maintaining flexibility for engineered structures in Greece (Kontogianni et al., 2013), and an application of robust decision making for planning coastal resilience for Louisiana in the United States (Groves and Sharon, 2013).

Flooding and water management risk

Climate change is projected to disrupt global and regional water cycles (IPCC, 2013). This leads to a number of potential risks, including more frequent and/or intense floods and changes to the water supply-demand balance, including potential water deficits (IPCC, 2014).

River, Surface Water and Urban Flooding Risks

Projections of future climate change suggest extreme precipitation events will become more intense and more frequent by the end of this century in many areas (IPCC, 2013). This has the potential to increase river and surface water floods (flash floods) (Kundzewicz et al., 2014). There are a broad range of adaptation options to address these risks (Wilby and Keenan, 2012), many of which are similar to coastal flooding. However, the analysis of adaptation costs and benefits is more challenging in this area, because of the probabilistic nature of flooding extremes and the high site-specificity. Earlier reviews found a low coverage of cost-benefit assessments on this topic, but this has expanded significantly in recent years.

At the global level, the World Bank's EACC study (2010) looked at the costs of flood protection, using a global hydrological model in an impact assessment. The study estimated the costs of adaptation at around USD 4-7 billion per year in the period 2010-50 for developing countries. At the regional, national and local scale, detailed hydrological models can be linked to probability-loss functions or depth damage functions to analyse adaptation costs and benefits. Such studies often assess the economic benefits of maintaining risk protection standards (e.g. maintaining a 1 in 100 year event under rising risks from climate change). These are then compared to the costs of flood protection, drawing on the costs of these programmes (e.g. Europe [HKV and RPA, 2014] and in the United States [MMC, 2005]). As an example, Rojas, Feyen and Watkiss (2013) estimated the economic benefits of adaptation from maintaining levels of river flood protection across Europe (EU27) (at a minimum 1 in 100 level) at EUR 8 billion per year by the 2020s, EUR 19 billion per year by the 2050s for a medium emission scenario (undiscounted). The authors concluded that these benefits were high compared to the likely costs of protection. There are similar studies at the national and river-basin level in many countries (e.g. in the Netherlands (Delta Committee, 2008; Bouwer et al., 2010) and the United Kingdom (Evans et al., 2004; Defra, 2011).

Such studies show that adaptation has potentially large benefits in reducing flood related damages under climate change. However, the scale of investment costs is also substantial, due to capital intensive flood defences and high maintenance costs. A similar finding emerges from investment and financial flow studies, which look at the likely increases in flood defence expenditure under climate change and find high additional costs (e.g. in Bangladesh [UNDP, 2011] and Nepal [IDS, 2014]) though also high benefits.

However, as with the coastal sector, many of these studies use impact assessment methodologies, and the same issues identified above therefore apply with respect to assumptions about foresight. As a consequence, adaptation is moving in a similar direction towards early low-regret options and consideration of uncertainty. There is some evidence on the costs and benefits of early low regret options for flood protection in the existing disaster risk reduction literature. Mechler et al. (2014) undertook a systematic review of the costs and benefits of flood risk management appraisals (*ex ante*) and evaluations (*ex post*), analysing 27 studies. The study found an average benefit to cost ratio of just under 5 to 1 for flood related risks. This review was further expanded in ECONADAPT (2015) which found options with high benefit to cost ratios include:

- Meteorological and hydrological information, forecasting and use in early warning systems (e.g. in the United States [EASPE, 2002; MMC, 2005], in Europe [IDRS, 2008; Desbartes, 2012; World Bank, 2013], and for developing countries [World Bank, 2012]).
- Disaster risk management and emergency, contingency and preparation response plans and awareness raising (Hawley et al., 2012), as well as creating the enabling environment for adaptation (Wilby and Keenan, 2012).
- Enhanced maintenance regimes for drainage and sewage systems (e.g. Moench et al., 2009; ECA, 2009; Ranger et al., 2011).
- Risk transfer including insurance, reserve funds and risk pools and risk facilities, especially for more extreme events (see Jongman et al., 2014, for an analysis in Europe).
- Household level adaptation responses that can either reduce risks or reduce damages (as shown by adaptation cost curves in the United Kingdom (ASC, 2011) and analysis of household level options in developing countries (e.g. World Bank, 2011).
- Integrated water resource management (e.g. Mechler, 2005) and climate smart agriculture (see later section).

Many of the most promising options are "behavioural" or soft measures – information and education, preparedness, forecasts and warning systems, emergency responses (see Hawley et al., 2012) – which are low-regret but not cost-free (Wilby and Keenan, 2012). There is evidence to suggest that the benefits of these options increase with greater levels of climate change (e.g. ECA, 2009), though on their own, there are limits (World Bank, 2011). There is also a greater focus – and evidence of higher benefit to cost ratios in developing countries – for community based interventions (see Moench et al., 2009). Mechler et al., 2014) highlights that there are a number of key assumptions and methodological challenges in such studies, and a key issue for the estimation of benefits is whether indirect and intangible effects are included.

As with coastal adaptation, there is also a move towards ecosystem-based and spatial options in a number of OECD countries. This includes spatial options that move beyond engineered control, such as the "room for the river" strategy in the Netherlands. These options include: watershed management including enhanced conservation and restoration, notably of upstream catchments; natural flood plain management, including water flow regulation and controlled flooding; and natural protection structures as an alternative to concrete. There has been a review of the costs and benefits of green schemes in Europe (HKV and RPA, 2014). This identifies studies on ecological variants (e.g. reed-land) of flood defences in the Netherlands (De Bel, Schomaker and van Herpen, 2011), wetland restoration in Stockholm (Kettunen, 2011), flood storage in the Humber estuary in the United Kingdom (EA, 2009c) and for the Elba in Germany (Teichmann and Berghöfer, 2010; TEEB DE, 2014). However, it is worth noting that benefits are often delivered in the future, due to the time for full ecosystem establishment (Naumann et al., 2011).

There has been less analysis of intra-urban flooding, though some country level studies consider adaptation costs (e.g. the United Kingdom [Evans et al., 2004] and

Germany [Tröltzsch et al., 2012]). There are also examples at the city scale. Desjarlais (2011) performed a cost-effectiveness analysis of urban water drainage in Montreal. There are also cost-benefit studies of sustainable urban drainage systems (RH DHV, 2012). Most recently, Copenhagen has developed and undertaken a cost-benefit analysis for a cloudburst plan (City of Copenhagen, 2012). All of these studies show potentially high adaptation benefits, but investment costs are often very high.

There are also examples of iterative adaptive management (e.g. the Delta Programme Kind, 2014; Eijgenraam et al., 2014), real option analysis (for water and flood risk infrastructure in the United Kingdom (Gersonius et al., 2013) and robust decision making to flood risk management (e.g. in Ho Chi Minh City in Viet Nam: Lempert et al., 2013).

Water supply and management risks

Water supply and wastewater services – and the sectors and activities that rely on them – are vulnerable to climate change. However, there is high uncertainty, making adaptation challenging. Adaptation to reduced water availability can include management of supply and demand. Supply measures include: increasing water storage capacity (e.g. the construction of dams or storage capacity, off-stream reservoirs, rainwater harvesting, artificial wetlands, off stream polders); improving water distribution (e.g. leakage control and meters); greywater reuse and rainwater harvesting; desalination; water transfer; aquifer storage and recovery; and water shipment. Demand measures involve increasing water use efficiency and reducing water consumption through changed sectoral activity, behavioural change, and technological uptake (e.g. water efficient appliances). Early reviews (OECD, 2008) found few studies in this sector, but more evidence has emerged in recent years.

At the global level, Kirshen (2007) estimated additional investment and financial flows for water supply at USD 9-11 billion per year in 2030. The EACC study (World Bank, 2009; Ward et al., 2010) estimated adaptation costs for developing countries at USD 10-11 billion per year between the period 2010-2050, based on the cost of meeting future water demand. Other aggregated estimates also exist. Hughes, Chinowsky and Strzepek (2010) estimated adaptation costs for water supply of 1–2% of baseline costs for all OECD countries, or about USD 5.5 billion per year.

There are also a number of studies at national level. These include studies in the Netherlands on climate proofing the water system (Van Ierland, 2006; De Bruin et al., 2009), and on water management in California (Tanaka et al., 2006). These studies report that adaptation costs could be high. There are also studies that use general equilibrium models to look at water adaptation costs, including Faust, Gonseth and Vielle (2012) in Switzerland, and analysis of network loss reductions by the Bank of Greece (2011). Metroeconomica (2006) estimated adaptation costs for anticipated water deficits in South-East England and South-East Scotland up to 2100, using indicative cost-curves and cost-effectiveness analysis, while the ASC (2011b) developed household water adaptation cost curves for the United Kingdom. Studies also exist on water management at the local scale. Anderson (2008) examined the economic benefits of water reuse in Sydney in the context of future water supply and demand. Mánez and Cerdà (2014) used a cost-benefit analysis to prioritise adaptation measures in Valencia and Catalonia. Skourtos, Kontogianni and Tourkolias (2013) developed an adaptation cost database for technologies for water saving for use at the European level.

There are also studies in developing countries (see ECONADAPT [2015] for further information). These include a broad geographical coverage, with studies in Central America (Bárcena, A. et al., 2010), South Africa (Callaway et al., 2006), Kenya (SEI, 2009), Ethiopia (World Bank, 2010), Ecuador (Vergara et al., 2007), Nepal (Dhakal and Dixit, 2013), China (Kirshen et al., 2005), Costa Rica, Dominican Republic, the Gambia, Bangladesh, Honduras, and Peru (UNDP, 2011), Jordan and the Maldives (UNFCCC, 2010).

There are some estimates of the costs of adapting wastewater and storm-water infrastructure, as well as water treatment costs, under climate change. These include studies in the United Kingdom on the costs for upgrading wastewater networks due to more frequent low-flows in rivers (ICF International, 2007) and cost-effectiveness analysis for agriculture and sewage treatment works to comply with the EU Water Framework Directive and Habitats Directives in the context of climate change (mitigation and adaptation) at sub-catchment level (Martin-Ortega et al., 2012); in Toronto, Canada on the costs of building new treatment plants, improving the efficiency of plants or increasing retention tanks (Dore and Burton, 2001); and in Boston on the costs of extra treatment of wastewater under climate change (Kirshen et al., 2004). In Sweden, the costs of increased infrastructure were estimated for wastewater plans to address water supply contamination from climate change risks and increased separation/inactivation of micro-organisms in water treatment plants (SCCV, 2007). Sussman et al. (2014) collated national and regional estimates for adapting water infrastructure in the United States.

There are several studies on the costs of adaptation for hydro-electricity, in terms of electricity system planning (using demand and energy optimisation models) as well as individual options for plants. Examples include studies in Brazil (Margulis and Dubeux, 2011), Ethiopia (World Bank, 2010) and Nepal (IDS, 2014). These indicate potentially large costs from the additional capacity needed to address demand shortfalls, though the outcomes vary significantly with climate projections. There are also some studies of the costs of adaptation in relation to the abstraction temperature of river water for cooling for thermal and nuclear power plants, an issue that emerged in the 2003 European heat wave, with estimates at European scale (Mima et al., 2011; CEPS/ZEW, 2010) and in some countries (e.g. in Germany [Tröltzsch et al., 2012]). Finally, there are some studies that consider adaptation costs for river transport, which is important on the major river systems of Europe. This includes analysis along the Rhine (Jonkeren, 2009) and other major river navigation routes (ECONET, 2014).

Recent discussion has moved towards low-regret adaptation options and the consideration of uncertainty. There are a set of early adaptation options that have high benefit to cost ratios (e.g. water efficiency measures, enhanced climate and hydrological monitoring and information [ECA, 2009; ASC, 2011]) as well as options that help improve watershed management (integrated water resource management and ecosystem based adaptation). There are also some examples of decision making under uncertainty, notably with robust decision making in California (Lempert and Groves, 2010), the Colorado River Basin (Groves et al., 2013) and for dam design in Greece (Nassopoulos et al., 2013), an application of real options analysis (Jeuland and Whittington, 2013) to water investment planning on the Blue Nile for large dams, and an application of decision pathways (iterative risk management) for water investment planning in London (Darch et al., 2011).

Other risks to infrastructure

While heavy precipitation and flood related damage are key risks to infrastructure, there are other climate related risks. At the global level, these were estimated using an investment and financial flow analysis in the UNFCCC study (2007), which estimated very high costs. Subsequently the World Bank's EACC study (2010) used an impact assessment method for adjustments in building standards for changing average temperature and precipitation, estimating infrastructure costs for developing countries at USD 13.5-27 billion per year in the period 2010-50, over half of which was for urban infrastructure.

More specific studies also exist. In the transport sector, there are some estimates of the additional cost of adaptation for transport infrastructure, including road and rail (Jochem and Schade, 2009; SCCV, 2007, in Sweden; Tröltzsch et al., 2012, in Germany). Recent studies in the United Kingdom indicate many of these risks can be addressed at low cost as part of planned maintenance and refurbishment regimes (Atkins, 2013) though high costs may be associated with strengthening bridges vulnerable to climate change (see Wright et al., 2012, in the United States). There have also been adaptation cost studies on road infrastructure in developing countries, including in Ethiopia and Ghana (World Bank, 2010).

In relation to colder regions, the costs of adaptation for infrastructure were considered for Alaska by Larsen et al. (2008), which reported that infrastructure costs could increase by as much as 10-20% in present value terms (USD 3.9-6.6 billion for the period 2006-30). Similarly, a cost study by Zhou et al. (2007) examines the infrastructure costs in the Northwest Territories of Canada. A number of studies have also considered the potential costs of adaptation of non-tropical windstorms, including Hunt and Anneboina (2011) in Europe, Tröltzsch et al. (2012) in Germany and SCCV (2007) in Sweden.

In recent years, the focus has moved towards climate risk screening for new infrastructure, as retrofitting of infrastructure is often expensive. By identifying risks at the outset, it is possible to avoid locating infrastructure in areas that are exposed to current or future climate risks. The consideration of future risks can also be used to change design, for instance, to build-in higher protection levels, increase flexibility or to allow more robustness. All these responses can build resilience to future risks. However, these can incur higher upfront costs, which need to be weighed against the expected benefits: As an example, the World Bank (2006) estimated that accounting for future climate in high-risk projects today could potentially increase project costs by 5-15%. While this may be justified for critical infrastructure (water supply, and health and emergency), it may not be justified in all cases, especially given the timing (and uncertainty) of future benefits and the (economic) lifetime of investments.

Agriculture

Agriculture is a highly climate sensitive sector and climate change has the potential for a large number of possible risks (IPCC, 2014). It may impact directly and indirectly on crop production, value chains and trade, with potentially negative effects (e.g. lower rainfall and increasing variability) but also positive effects (e.g. CO_2 fertilisation and extended growing seasons). There are also potential impacts from changes in extremes, and the range and prevalence of pests and disease. Similar issues also arise for horticulture, viticulture, industrial crops and livestock. While negative impacts on yield are projected for most crops in tropical and temperate regions above 2 °C, the earlier patterns are uncertain, and include potential gains as well as losses in yields, with strong variations between crops and across regions (Porter et al., 2014).

There are a number of potential adaptation options to address these risks (Ignaciuk and Mason-D'Croz, 2014), though information on the costs and benefits of these varies significantly (not least due to the uncertainty over future impacts). These include a wide range of options, such as changing planting dates, use of new varieties, diversification and sustainable soil and water management techniques. OECD (2008) identified a good coverage of adaptation benefits for agriculture, from two sets of studies. The first were based on autonomous (farm-driven) adaptation using crop models and impact-assessment (e.g. Parry et al., 2004). These generally consider the increased use of irrigation and fertiliser to address failing yields. The results can be used as part of or as an input to global models, which allow autonomous market adaptation from trade, taking account of the total impact of climate change rather than just the direct domestic impacts. The second were based on econometric (Ricardian) analysis (e.g. Seo et al., 2009), to assess the relationship between climatic factors and land value or farm net revenues.

At the global level, a UNFCCC study (McCarl, 2007) estimated adaptation costs for the agricultural sector (research, extension and irrigation) at USD 14 billion per year globally by 2030, of which 50% was in developing countries. The International Food Policy Research Institute (IFPRI) (2009) – as part of the EACC study – used an agricultural supply-and-demand projection and a biophysical crop model to estimate agricultural productivity investments and adaptation costs. For developing countries, EACC (2010) estimated costs of USD 2.5-3 billion per year. These studies found lower crop yields with climate change (especially for irrigated and rain-fed wheat and irrigated rice) but found costs of adaptation were low, because welfare was restored through trade, with some regions and countries becoming major food importers.

A large number of similar adaptation studies have been undertaken in developing countries (see ECONADAPT [2015] for further information). As an example of the coverage across countries, these include crop modelling studies (in Bangladesh, Bolivia, Ethiopia, Ghana, Mozambique, Samoa and Viet Nam [World Bank, 2010b, c, d, e, f, g], India [Markandya and Mishra, 2010] and Brazil [Margulis and Dubeux, 2010]), as well as sector investment and financial flow analysis in Bangladesh, Colombia, Ecuador, the Gambia, Liberia, Namibia, Niger, Paraguay, Peru, Togo, Turkmenistan and Uruguay (UNDP, 2011).

Analysis of the earlier crop modelling impact-assessment based studies compared to more recent policy studies reveals some insights. Earlier studies focus on a narrow set of adaptation options, particularly irrigation and fertiliser use. These studies rarely consider constraints and they exclude cross-sectoral issues. Studies that consider such factors (e.g. Iglesias et al., 2012, for Europe) find either policy constraints (e.g. on fertiliser use) or higher costs (e.g. from increasing competition for water in areas of water scarcity) compared to earlier studies. They also have optimistic assumptions about the substitution of domestic production for international trade under climate change, ignoring the costs that would be borne by local farmers as part of this transition, as well as the externalities associated with potentially lower food security levels. Finally, they do not consider uncertainty, considering scenarios one at a time, and assume high capacity and future foresight at farm level.

Recognising these issues, more recent studies have shifted to more immediate timescales and focused on a wider set of practical adaptation options. They have also started to consider decision making under uncertainty. Much of the recent focus in the literature has been on climate smart agriculture (FAO, 2013), which encompasses sustainable agricultural land management practices such as agroforestry, soil and water conservation, reduced or zero tillage, and use of cover crop. These options improve soil water infiltration and holding capacity, as well as nutrient supply and soil biodiversity. They also reduce risks from rainfall variability and soil erosion, increase soil organic matter, soil fertility and productivity, as well as reduce greenhouse gas emissions by reducing soil emissions.

There has been analysis of the costs and benefits of climate smart options. Examples include qualitative cost-benefit assessment of various options in Canada (British Columbia, 2013), as well as a cost-benefit analysis of conservative or low tillage in Germany (Tröltzsch et al., 2012), though the latter found modest ratios of benefits to cost. These options are particularly attractive in developing countries, due to the large benefits associated with rain-fed agriculture, and there are studies on their costs (McCarthy, Lipper and Branca, 2011), benefits (Branca, 2011) and cost-benefit analysis (e.g. Branca et al., 2012; ECA, 2009; Lunduka, 2013). These studies report high benefits for these options, reduced greenhouse gas emissions, and in some cases large co-benefits. However, McCarthy, Lipper and Branca (2011) highlight that there is a high variation in costs between sites, and also, that many of the adaptation options have high opportunity or transaction costs, since introduction involves labour and land costs as well as foregone crop income. These costs are a barrier to adoption of climate smart options, particularly in subsistence economies. Benefit to cost ratios also vary with discount rate (as does the rate of return), as some options take several years to establish benefits, while costs are immediate. Furthermore some of the benefits of these options (e.g. environmental improvements or greenhouse gas emission reductions) may not accrue to local farmers. These various issues highlight the need for planned support.

There has also been a focus on other early low-regret options. In OECD countries, there has been work to identify such options in the agricultural sector (Wreford and Renwick, 2012; Moran et al., 2013). Promising options identified in such studies include increasing water supply through on-farm storage reservoirs and incentivising efficient water management, the introduction of soil conservation measures and increasing spend on research and development. The recent analysis by the Intergovernmental Panel on Climate Change (IPCC) reported that some adaptation approaches (e.g. cultivar adaptation and planting date adjustment, in combination with other measures) are on average more effective than irrigation and fertiliser optimisation (Porter et al., 2014). A study in Germany (Tröltzsch et al., 2012) found that crop switching was a promising adaptation option, with high benefit to cost ratios. There are also studies that look at agriculture and irrigation in areas of water scarcity, as outlined in the earlier section on water supply and management risks. Notable studies include the early work in Australia (Howden et al., 2003), which highlighted the high benefit to cost ratios of research and development to improve the evidence base, and the more recent focus on vulnerable areas such as the Murray-Darling Basin (Adamson et al., 2009; Conor et al., 2009). The latter identifies low cost adaptation strategies for early moderate changes in water availability, such as irrigation efficiency and water allocation management.

There are also now more sophisticated national, regional and global assessments being undertaken, which are considering global food markets, trade and the cost of climate change adaptation (FAO, 2015). These include studies that link crop models and global trade models (e.g. using recursively dynamic partial equilibrium models), using these to explore climate change impacts and adaptation policies including consumer support policy (e.g. Mosnier et al., 2014, in four East Asian countries). Such studies highlight that looking only at crop yield projections in one region is inadequate to derive conclusions on climate change impacts and adaptation. More recent studies have also factored in uncertainty and robustness to such global assessments and considered transformational adaptation (e.g. Leclère et al., 2014), including uncertainty with stochastic modelling (Fuss et al., 2011; 2015) to see how this affects strategies and costs, as well as expanding the list of options to include climate smart agriculture.

In developing countries, many of the emerging low-regret adaptation options that are cited overlap with existing agricultural development strategies. While this raises some issues of attribution, in relation to the overlap with existing agricultural development, they do have high benefit to cost ratios. Promising options include (Ranger and Garbett-Shiels, 2012; ECONADAPT, 2015): climate information, agro-meteorological information, seasonal forecasting and early warning; research and development; crop switching/planting and diversification (agronomic management); pest and disease management, including post-harvest losses; soil and water management; ecosystem based adaptation; and insurance. Some countries have assessed and costed these options in sector adaptation plans (e.g. Government of Tanzania, 2014). Such options often work best when implemented as portfolios, rather than as single solutions, as found by Di Falco and Veronesi (2012). There are also some studies which consider (and cost) agricultural options using iterative adaptive management planning (e.g. Downing et al., 2011; Matiya, Lunduka and Sikwese, 2011).

Finally, there are some examples of iterative adaptive management in the agricultural sector. Examples include the United Kingdom study on the Economics of Climate Resilience (Frontier Economics, 2013) which developed adaptation pathways (roadmaps) for the sector. This identified early options that focused on building the enabling environment and information for adaptation in the farming sector, rather than technical options. Further examples include the application of iterative management in the Ethiopian Climate Resilience Strategy (FRDE, 2014) and real options analysis to agricultural irrigation in Mexico (World Bank, 2009).

The evidence base on adaptation costs and benefits for horticulture, viniculture, livestock, forestry and fisheries (including aquaculture) is less developed, though some studies are emerging, for example in relation to forestry management and fire control (e.g. Price et al., 2012; Khabarov et al., 2014) and viniculture (Zhu et al., 2013), both of which are priority areas for early adaptation given the long life-cycles for production.

Heat-related and extremes – health, energy and the built environment

Climate change will lead to more frequent high temperature extremes, and heat waves will occur with a higher frequency and duration (IPCC, 2013), this will increase health impacts though there will also be potential health benefits from the reductions in cold related impacts due to warming. Higher temperatures – both average and heat extremes – will also affect building comfort and energy demand for heating and cooling. The adaptation options for these two risks are closely related and are discussed below.

Health adaptation

Heat wave early warning systems (heat alert) are an early low-regret option to address heat related mortality and morbidity. The IPCC (Smith et al., 2014) reviewed studies that considered the effectiveness of such systems, reporting most schemes led to fewer deaths during heat waves. There is also *ex post* data on the costs of these schemes, e.g. in France

(ONERC, 2009) and Europe (WHO, 2009). The estimated costs range from under EUR 1 million to up to around EUR 10 million per scheme, depending on the cost categories included, with upper estimates including costs of additional medical personal and/or resource costs.

There are studies that have assessed the costs and benefits of these systems in relation to current risks (e.g. Ebi et al., [2004] for Philadelphia, the United States) and for future climate change (e.g. Hunt and Watkiss [2010] for London and Tröltzsch et al. [2012] in Germany). These indicate they are a cost-effective response (i.e. with high ratios of benefits to costs) though it is noted that the future annual costs of these schemes will rise as the systems are triggered more frequently with future climate change (though benefits will also increase). However, additional adaptation to address heat-related mortality is likely to be needed, as there are residual deaths during heat extremes even when these systems are in place. This is likely to require more expensive options. As an example, Michelon, Magne and Simon-Delaville (2005) report that immediately following the 2003 heatwave, EUR 150 million was invested for additional staff and cool rooms in elderly residential homes in France. There have also been cost-benefit assessments of increased cooling in hospitals (Tröltzsch et al., 2012). Most health adaptation plans therefore stress the need for future intervention from outside public health, notably the built environment.

Built environment (energy demand)

Climate change will affect future energy demand, increasing summer cooling, but reducing winter heating. These responses are largely autonomous, and can be considered as an impact or an adaptation, though they are strongly influenced by other socio-economic factors, notably income. While energy demand for cooling will increase in warmer countries, in the OECD this will be driven by temperature, but in developing countries it will be dominated by rising incomes (Arent et al., 2014). Isaac and van Vuuren (2009) estimate large increases in energy related cooling demand in Asia from climate change, and increases are projected to be especially high in India (Akpinar-Ferrand and Singh, 2010).

There are a growing number of studies that assess costs of cooling demand. Mima, Criqui and Watkiss (2011) assessed cooling and air conditioning costs for Europe (by region), and for the United States, China and India using a least cost-optimisation energy model, but also considered the additional (discounted) cost of air conditioning units. In the European Union, cooling costs were estimated at EUR 30 billion per year by 2050, undiscounted, and three times this amount by 2100, with a strong distributional pattern and higher costs in the South. The costs of air conditioning demand in India were estimated to be extremely high, at several hundred billion USD per year (undiscounted) by the end of the century. There are similar studies at the national level (e.g. in the United Kingdom [Defra, 2011] and in Spain [Pilli-Sihlova et al., 2010]). In countries where several studies exist, these reveal a wide range of costs. Sussman et al (2014) report on five national studies in the United States, indicating annual costs of USD 6 billion to USD 87 billion (undiscounted). These studies show that the autonomous costs of increased cooling could be large, even if these are offset by reductions in heating, as in many OECD countries. If this cooling is delivered with air conditioning, this could contribute to higher greenhouse gas emissions, conflicting with mitigation objectives (unless electricity is decarbonised). However, there is the potential for air conditioning to reduce health impacts as a co-benefit (see Ostro et al., 2010).

Recent studies have therefore focused on alternatives to air conditioning. There are studies at national and local level on the costs of passive and retrofit options, particularly in Europe (e.g. van Ierland et al., 2006; Arup, 2008; ASC, 2011; Mima, Criqui and Watkiss, 2011).

These include a range of options such as simple shading and orientation, design and building codes, and low- and very low-energy consumption buildings. While these apply primarily for new buildings, some also consider retrofitting of existing houses. These assessments find the benefits and costs to vary strongly across the range of climate projections, and with the assumptions underpinning the assessments. A general finding, however, is that it is more expensive to retrofit existing houses than to include these measures in new buildings. There are also challenges related to implementation (see Neufeld et al., 2010), which are likely to lead to policy costs. An alternative or complementary option is to reduce urban over-heating using spatial planning, such as green spaces and open plan development. There are some studies that look at the benefits of these schemes, though in the OECD context, the costs are very high, because of the costs of land-use change.

As well as over-heating, some recent studies have started to consider multiple urban risks and cross-sectoral responses. For example, a study by Pohl et al. (2014) undertook a cost-benefit analysis of a set of adaptation options to a range of risks in a part of Rotterdam (the Netherlands), including heat, storm water flooding and drought, finding highest benefit to cost ratios for awareness raising and behavioural change. A particular option that has been advanced for such cross-sectoral studies has been green roofs, and a number of economic studies exist (van Ierland et al., 2006 in the Netherlands; LCCP, 2009 in London; Tröltzsch et al., 2012 in Düsseldorf; Nurmi et al., 2013, in Finland). These schemes offer multiple co-benefits (e.g. reduced energy, storm-water management, sewer overflow, air quality, urban heat island and greenhouse gases), though the literature often reports modest benefit to cost ratios.

Other health risks

There are a number of other potential health impacts from climate change, both direct and indirect, including: water, food and vector-borne disease; deaths, injuries and mental well-being from extreme events; changes in air pollution and allergens; effects from altered agricultural production and food insecurity, and conflict. There are also risks to health infrastructure and to occupational health. At the global level, Ebi (2008) estimated the costs of adaptation to diarrhoeal disease, malnutrition and malaria using unit prevention costs at USD 5 billion per year by 2030 in developing countries. Parry et al. (2009) argued that these were significant underestimates, as they only included 30-50% of extra disease burden, and excluded additional costs relating to public health infrastructure. The EACC study (2010) used a similar approach, using preventative costs for adaptation for malaria and diarrhoea for developing countries up to 2050, but took account of declines in the baseline incidence due to development in the future. It estimated adaptation costs at only USD 1.5-2 billion per year in developing countries, with most of these arising in Africa.

A number of other studies have looked at health adaptation costs at the national level, though the focus has been on developing countries (see ECONADAPT, 2015) with national adaptation cost studies in India (Chiabai et al., 2010), Paraguay (UNDP, 2011) Saint Lucia (ECLAC, 2011c) and Ghana (UNFCCC, 2010), as well as adaptation costs studies in Kenya for malaria (SEI, 2009) and water borne diseases in Tanzania (EC, 2009). A common theme from these studies is that, in the near-term, the most effective measures in developing countries are programmes that implement and improve basic public health measures such as clean water and sanitation, essential health care including vaccination and child health services, disaster preparedness and response, and alleviate poverty (Smith et al, 2014) as well as enhanced surveillance and monitoring. These options have demonstrated high benefit to cost ratios (see Hunt [2011] for water and sanitation options).

However, for some risks, especially in OECD countries, costs are likely to be higher. This includes the potential costs of addressing risks to water and waste-water infrastructure as well as water quality in OECD countries (see above). It also includes the potential costs of large-scale vaccination programmes, such as against tick borne disease (see Hsai et al., 2002; Desjeux et al., 2005). Finally, it includes the potential adaptation costs to address air pollution related risks from climate change, particular for ozone. Epstein and Mills (2005) and Liao et al. (2010) both report high costs in the United States – either from increased treatment costs or pollution control – to address increased asthma cases due to climate change.

Overall, while the evidence base has increased, this is still an area with major gaps, and most analysis is focused on options that have easily measurable costs attached. The cost coverage is also incomplete: capital costs are often neglected, as are resource and policy costs. There are, however, some initiatives which are starting to address these gaps, notably a recent tool developed by the World Health Organisation (2013), to aid decision makers in making estimates of health adaptation costs.

Biodiversity and ecosystem services

Climate change poses potentially large risks to terrestrial, aquatic and marine biodiversity and the ecosystem services they provide (provisioning, regulating, cultural and supporting services). It will shift geographic ranges, seasonal activities, migration patterns, abundances, and species interactions, and has the potential to increase species extinction (Settele et al., 2014). Previous reviews have identified a major gap in this area, reflecting the challenges involved in quantification and valuation. The literature that does exist focuses on the costs of protection and restoration of habitats and species, though there is literature on ecosystem based adaptation ("green" options) discussed in earlier sections (see coastal and floods risk sections).

Earlier studies (Berry, 2007) estimated the global costs of establishing and managing protected areas under climate change at USD 36-65 billion per year by 2030, noting the costs would be as high as USD 290 billion per year when extended to conservation of the wider matrix of landscapes. Some studies at national level also exist. Berry et al. (2006) estimated the adaptation costs (for restoration and re-creation) in the United Kingdom for a number of habitats. The UNDP investment and financial flow analysis of Costa Rica estimated the costs of adaptation for the biodiversity sector (i.e. conservation of terrestrial, marine and aquatic ecosystems, prevention of forest fires, and awareness raising) at USD 60 million per year in 2015 rising to USD 76 million per year in 2030, i.e. a total of USD 1.3 billion over the period (UNDP, 2011). A similar study in Peru (UNDP, 2011) estimated adaptation costs for fisheries at USD 0.78 billion to 2030 (i.e. approximately USD 40 million per year). Van Ierland et al. (2006) estimated the costs of establishing a national ecologic network, in the Netherlands and additional adaptation under climate change at EUR 135 million per year.

Cartwright (2013) analyses adaptation measures in a metropolitan region in Durban, South Africa, finding positive benefit-cost ratios for the three ecosystem-related measures. While most early low-regret options centre on the reinforcement or enlargement of existing measures to protect biodiversity (e.g. use of protected areas, buffer zones, ecological corridors, reducing habitat fragmentation), there are some new approaches (e.g. selection of species, translocation of species, management of alien species), alongside enhanced information and monitoring. However, there is very little evidence on the costs and benefits of these options: one study in Finland analysed the conservation of grassland butterflies under a changing climate (Tainio et al., 2014) finding that buffer zones were most cost-effective while the costs of translocation were relatively modest compared to dispersal corridors.

Nonetheless, in recent years, more literature on the value of ecosystem services has emerged (TEEB, 2009; TEEB 2010) that provides potential inputs for the analysis of adaptation costs and benefits. These studies highlight the economic values of restoration projects as an adaptation measure, and have assessed the benefit to cost ratio for restoration of different biomes and ecosystem, finding high benefit to cost ratios, especially for grassland, tropical forests, wood- and shrub-land, and mangroves.

There is also an increased interest in the application of adaptive management to this sector, though studies to date have not focused on economics. It has also proved challenging to apply the new economic tools for decision making under uncertainty to this area. The only study identified for this analysis is an application of portfolio analysis to investigate genetic material that could be used for the restoration or regeneration of forests under climate change futures (Crowe and Parker, 2008).

A key conclusion is that the evidence base remains limited, and the information that does exist is difficult to transfer, due to the complexity of estimating the impacts of climate change, and the additional challenge of valuation. The studies that do exist indicate that aggregate costs could be high, and that there are potential opportunity and policy costs. This sector remains a priority for research.

Business, services and industry

Business, services and industry are the main source of economic activity in OECD countries, yet the understanding of risks to these sectors remains limited. There are some studies of potential effects of climate change on tourism (Hamilton et al., 2005). For beach tourism, the impacts of climate change are often assessed using a Tourist Climatic Index, and costed by using tourism expenditures. As an example, Amelung and Moreno (2012) find that in Europe there is a strong redistribution of tourism (and expenditures) away from southern countries such as Spain, Greece and Croatia (due to the increased heat in key summer months), but with positive effects on the climate for tourism in northern countries, such as the United Kingdom, Ireland, Germany, the Netherlands and Austria. These changes can be seen as an impact or an autonomous adaptation response. There is less literature on planned adaptation responses, though the Dominican Republic (UNDP, 2011) undertook an investment and financial flow analysis for tourism and estimated adaptation costs would rise from USD 16 million per year in 2015 to USD 57 million per year by 2030 (totalling USD 0.7 billion).

There are also several studies that look at winter tourism. For example, OECD (2007) assessed the costs of adaptation in the Alps, and the costs of additional snow machines and increased use to cope with decreased snow reliability in the lower altitudes ski resorts, as well as extending ski areas to higher elevations. There are also some studies on the cost of preparing slopes in the German region of Bavaria and adapting for summer tourism (cycling) (Tröltzsch et al., 2012).

However, the consideration of other areas remains low. There are a small number of studies that, for example, analyse the costs and benefits of information and avoiding heat induced productivity reduction for industry in Germany (Tröltzsch et al., 2012). There are also some studies on occupational health and workplace productivity (see the health and built environment sections above). Some studies are emerging on the potential economic opportunities from adaptation, including studies at the European (Triple E., 2014), country (e.g. BIS [2013] for the United Kingdom) and city level (KMatrix [2014] for London).

Finally, an important issue is the wider macroeconomic effects of climate change and adaptation, i.e. how effects in one sector (and in aggregate) cascade across the economy. For example, the studies reported above are primarily sector based assessments, though there are some examples of partial equilibrium analysis (e.g. in agriculture and energy). There are some early examples using computable general equilibrium (CGE) modelling to explore these issues. Bosello et al. (2012) used a CGE model to assess the wider economic costs of sectoral costs of adaptation for the coastal sector in Europe. Carraro and Sgobbi (2008) moved to the national level, and assessed the economic value of the impacts of climate change for economic sectors and regions, aggregated to provide a macroeconomic estimate (GDP) using a CGE model, and included autonomous adaptation induced by changes in relative prices and in stocks of natural and economic resources, as well as international trade effects (changes in prices inducing changes in production and demand). The Bank of Greece (2011) study used a general equilibrium model to estimate the macroeconomic cost of planned adaptation measures for the sectors of water, forests, transport, tourism, the built environment and coastal zones. A key priority is to extend these assessments to consider more detailed (and iterative) planned adaptation.

It is highlighted that this area represents a considerable gap in the knowledge of the potential impacts of climate change in OECD countries and potential adaptation costs. This includes the effects on business disruption, from direct and cross-sectoral impacts, including along supply chains.

Discussion of the current state of evidence and key gaps

An analysis of the evidence above reveals a number of key insights. Most of these estimates are from grey literature – only 25% are academic peer-reviewed articles (ECONADAPT, 2015). Moreover, most of the evidence is based on classic scenario-based impact assessment methods. This means that the majority of the studies are theoretical, focus on technical adaptation, and ignore uncertainty. Earlier studies show that adaptation has very high benefit to cost ratios and potentially low costs, though more recent studies indicate they are probably over-optimistic. As the evidence base in this area is still emerging, there is an urgent need for more empirical studies, to address key gaps, and to ensure information and lessons can be shared.

Perhaps most interestingly, there has been a major shift in the evidence base over the past few years, and this provides a number of key insights. First, more recent studies focus on early adaptation and low-regret options. That is, they identify different early options (with more focus on adaptive capacity, the valuation of information and soft options). Many of these early low-regret options will have lower costs than engineering based options (Agrawala et al., 2011a), and they often offer wider co-benefits. However, they are only the initial steps in a longer adaptation pathway, and are introduced early in the planning process, at a time when classic impact assessment studies induce very little action.

Second, more recent studies are more grounded in existing sectoral policy. Such studies identify that many adaptation options will have important opportunity, transaction or policy costs (DFID, 2014), which are not included in the earlier technical studies. These costs arise even for low-regret options such as climate smart agriculture or ecosystem-based adaptation. Experience from the mitigation domain has demonstrated that it is rarely as easy or cheap to implement low or no regret options as expected, due to a range of economic, information and policy barriers. There is also an increasing recognition that implementation costs in developing countries will need to consider existing development and governance challenges, which are likely to affect the effectiveness of adaptation or the costs of delivering options.

Finally, more recent studies frame medium and longer-term adaptation in a different way, using iterative risk management or decision making under uncertainty. The methods themselves are therefore different to the older studies, and they identify different options as a result. These approaches provide high potential benefits, using adaptive management to avoid future inefficient or ineffective adaptation. However, these approaches require higher adaptive capacity to implement than earlier studies.

Notes

1. The ECONADAPT project is funded by the European Union's Seventh Framework Programme for research, technological development and demonstration under grant agreement no 603906. The views expressed in this publication are the sole responsibility of the authors and do not necessarily reflect the views of the European Commission. The European Community is not liable for any use made of this information.
2. Co-funding was provided by: i) UK Department for International Development, as part of the project "Early Value-for-Money Adaptation: Delivering VfM Adaptation using Iterative Frameworks and Low-Regret Options" – this project has been funded by UK aid from the UK government; however the views expressed do not necessarily reflect the UK government's official policies: ii) Canada's International Development Research Centre (IDRC), as part of the project "The Economics of Adaptation and Climate-Resilient Development" – however the views expressed are entirely those of the study team and do not necessarily reflect the views of IDRC.

References

Ackerman, F. and E.A. Stanton (2008), *The Cost of Climate Change: What We'll Pay if Global Warming Continues Unchecked*, The Natural Resources Defense Counsil, New York.

Adamson, D., T. Mallawaarachchi and J. Quiggin (2009), *Declining inflows and more frequent droughts in the Murray-Darling Basin: Climate change, impacts and adaptation*, Australian Journal of Agricultural and Resource Economics, Vol. 53(1), *http://dx.doi.org/10.1111/j.1467-8489.2009.00451.x*.

ADB (2014), *The Economics of Climate Change in South Asia*, Asian Development Bank, Manila.

ADB (2009), *The Economics of Climate Change in Southeast Asia: A Regional Review*, Asian Development Bank, Manila.

ADB (2005), *Climate Proofing: A Risk-based Approach to Adaptation*, Asian Development Bank, Manila.

Adger, W.N. et al. (2007), *Assessment of adaptation practices, options, constraints and capacity. Climate Change 2007: Impacts, Adaptation and Vulnerability*, Contribution of Working Group II to the Fourth Assessment Report of the Intergovernmental Panel on Climate Change, M.L. Parry, O.F. Canziani, J.P. Palutikof, P.J. van der Linden and C.E. Hanson, Eds., Cambridge University Press, Cambridge, UK, 717-743.

Aerts, J.C. et al. (2014), *Evaluating Flood Resilience Strategies for Coastal Megacities*, Science, Vol. 344(6183), *http://dx.doi.org/10.1126/science.1248222*.

Aerts, J.C. et al. (2013), *Cost estimates for flood resilience and protection strategies in New York City*, Annals of the New York Academy Sciences, Vol. 1294, *http://dx.doi.org/10.1111/nyas.12200*.

AfDB (2011), *The Cost of Adaptation to Climate Change in Africa*, African Development Bank, Tunis.

Agrawala, S. et al. (2011a), *Adapting to Climate Change: Costs, Benefits, and Modelling Approaches*, International Review of Environmental and Resource Economics, Volume 5(3), *http://dx.doi.org/10.1787/5km975m3d5hb-en*.

Agrawala, S. et al. (2011b), *Plan or React? Analysis of adaptation costs and benefits using integrated assessment models*, Climate Change Economics, Vol. 2(3), *http://dx.doi.org/10.1142/S2010007811000267*.

Akpinar-Ferrand, E. and A. Singh (2010), *Modeling increased demand of energy for air conditioners and consequent CO_2 emissions to minimize health risks due to climate change in India*, Environmental Science and Policy, Volume 13(8), *http://dx.doi.org/10.1016/j.envsci.2010.09.009*.

AIACC (Assessments of Impacts and Adaptations to Climate Change) (2006), *Estimating and Comparing Costs and Benefits of Adaptation Projects: Case Studies in South Africa and The Gambia*, report submitted to Assessments of Impacts and Adaptations to Climate Change, the International START Secretariat, Washington, DC.

Amelung, B and A. Moreno (2012), *Costing the impact of climate change on tourism in Europe: results of the PESETA project*, Climatic Change *http://dx.doi.org/10.1007/s10584-011-0341-0*.

Anderson, J.M. (2008), *Adapting to climate change with water savings and water reuse*, Water Practice & Technology, Volume 3(2), *http://dx.doi.org/10.2166/WPT.2008036*.

Arent, D.J. et al. (2014), Key economic sectors and services. In: *Climate Change 2014: Impacts, Adaptation, and Vulnerability. Part A: Global and Sectoral Aspects.* Contribution of Working Group II to the Fifth Assessment Report of the Intergovernmental Panel on Climate Change. Cambridge University Press, Cambridge, United Kingdom and New York, NY, USA.

Arup (2008), *Your home in a changing climate: Retrofitting existing homes for climate change impacts*, Report for the Three Regions Climate Change Group, Greater London Authority, London.

ASC (Adaptation Sub Committee) (2011), *Research to identify potential low-regrets adaptation options to climate change in the residential buildings sector*, a report prepared by Davis Langdon for the Adaptation Sub Committee.

ASC (Adaptation Sub Committee) (2014), *Managing climate risks to well-being and the economy*, Adaptation Sub Committee Progress report 2014.

Atkins (2013), *Economics of climate change adaptation and risks*, a report prepared for the Highways Agency.

Bank of Greece (2011), *The environmental, economic, and social impacts of climate change in Greece*, Climate Change Impacts Study Committee, Athens.

Bárcena, A. et al. (2010), *The Economics of Climate Change in Central America*, ECLAC Subregional Headquarters in Mexico, Mexico City and Centeral American Integration System (SICA), El Salvador.

BIS (2013), *Adaptation and resilience (climate change)* (A&RCC), Published by Department for Business, Innovation & Skills, Department for Environment, Food & Rural Affairs and Department of Energy & Climate Change.

Bosello, F., C. Carraro and E. De Cian (2009), *An Analysis of Adaptation as a Response to Climate Change*, Working Paper, No. 26, Department of Economics Ca' Fosari University of Venice.

Bosello, F., C. Carraro and E. De Cian (2013), *Adaptation Can Help Mitigation: An Integrated Approach to Post 2012 Climate Policy*, Environment and Development Economics, Vol. 18(3). *http://dx.doi.org/10.1017/S1355770X13000132*.

Bosello F. et al. (2012), *Economic impacts of climate change in Europe: sea-level rise*, Climatic Change, Vol. 112(1), *http://dx.doi.org/10.1007/s10584-011-0340-1*.

Boyd, R. and A. Hunt (2004), *Costing the impacts of climate change in the UK: Overview of guidelines*, UKCIP Technical Report, UKCIP, Oxford, July 2004

Berry, P. (2007), *Adaptation options on natural ecosystems*, a report for the UNFCCC Secretariat, Financial and Technical Support Division.

Berry, P., A. Hunt and P. Nunes (2006), Chapter 6: Biodiversity in *Task 3 Report – Climate change impacts and adaptation: Cross-regional research programme, Project E – Quantify the cost of future impacts*, report by Metroeconomica Ltd. for Defra.

Bölscher, T. et al. (2013), *Adaptation Turning Points in River Restoration? The Rhine Salmon Case*, Sustainability, Volume 5(6), *http://dx.doi.org/10.3390/su5062288*.

Bouwer, L.M., P. Bubeck and J.C.J.H. Aerts (2010), *Changes in future flood risk due to climate and development in a Dutch polder area*, Global Environmental Change, Volume 20(3), *http://dx.doi.org/10.1016/j.gloenvcha.2010.04.002*.

Branca, G., L. Lipper and A. Sorrentino (2012), *Benefit-costs analysis of climate-related agricultural investments in Africa: A case study*, Italian Association of Agricultural and Applied Economics, Congress Papers, 2012 First Congress, June 4-5, 2012, Trento, Italy.

Branca, G. et al. (2011), *Climate-Smart Agriculture: A Synthesis of Empirical Evidence of Food Security and Mitigation Benefits from Improved Cropland Management*, Mitigation Of Climate Change In Agriculture Series, No. 3, FAO, Rome.

British Columbia (2013), *Farm Practices & Adaptation*, Climate Action Initiative, *www.bcagclimateaction.ca/adapt/farm-practices/*.

Broekx, S. et al. (2011), *Designing a long-term flood risk management plan for the Scheldt estuary using a risk-based approach*, Natural Hazards, Volume 57(2), *http://dx.doi.org/10.1007/s11069-010-9610-x*.

Brown S. et al. (2011), *Sea-Level Rise: The Impacts and Economic Costs of Sea-Level Rise on Coastal Zones in the EU and the Costs and Benefits of Adaptation*, in Watkiss, P. (ed.), The ClimateCost Project, Vol. 1, Stockholm Environment Institute, Sweden.

Brown, S., A.S. Kebede and R.J. Nicholls (2009), *Sea-Level Rise and Impacts in Africa, 2000 to 2100*, report prepared for the Stockholm Environment Institute.

Bruner, A., R. Naidoo and A. Balmford (2008), *Review on the economics of biodiversity loss: Scoping the science*, Review of the costs of conservation and priorities for action, European Commission, Brussels.

Burton, I. (2004), *Climate Change and the Adaptation Deficit*, in French, A. et al. (eds.), *Climate Change: Building the Adaptive Capacity*, Meteorological Service of Canada, Environment Canada.

Callaway, J.M. et al. (2006), *The Berg River Dynamic Spatial Equilibrium Model: A New Tool for Assessing the Benefits and Costs of Alternatives for Coping With Water Demand Growth, Climate Variability, and Climate Change in the Western Cape*, AIACC Working Paper, No. 31, The AIACC Project Office, International START Secretariat, Washington, DC.

Carraro C. and A. Sgobbi (2008), *Climate Change Impacts and Adaptation Strategies in Italy: An Economic Assessment*, Available at: *http://ageconsearch.umn.edu/bitstream/6373/2/dp080006.pdf* (accessed March 2015).

Cartwright, A. et al. (2013), *Economics of climate change adaptation at the local scale under conditions of uncertainty and resource constraints: the case of Durban, South Africa*", Environment and Urbanization, No. 6, *http://dx.doi.org/10.1177/0956247813477814*.

Chiabai A, Balakrishnan S, Sarangi G, and Nischal S (2010). *Human Health*. In Markandya A and Mishra A (Eds). Costing Adaptation: Preparing for Climate Change in India. TERI Press: New Delhi, India. ISBN 978-81-7993-388-6.

CCRIF (Caribbean Catastrophe Risk Insurance Facility) (2010), *Enhancing the ClimateRrisk and Adaptation Fact Base for the Caribbean: An informational brochure highlighting the preliminary results of the ECA Study*, CCRIF's Economics of Climate Adaptation Initiative, Cayman Islands.

CEPS and ZEW (2010), *The Fiscal Implications Of Climate Change Adaptation*, Centre for European Policy Studies and Centre for European Economic Research, Brussels and Mannheim.

Chambwera, M. et al. (2014), *Economics of adaptation. In: Climate Change 2014: Impacts, Adaptation, and Vulnerability. Part A: Global and Sectoral Aspects*. Contribution of Working Group II to the Fifth Assessment Report of the Intergovernmental Panel on Climate Change. Cambridge University Press, Cambridge, United Kingdom and New York, NY, USA, pp. 945-977.

Ciscar, J.-C. et al., (2011), *Physical and Economic Consequences of Climate Change in Europe*, PNAS, (Proceedings of the National Academy of Sciences), Vol. 108(7), *http://dx.doi.org/10.1073/pnas.1011612108*.

City of Copenhagen (2012), *Københavns Kommunes Skybrudsplan 2012*, City of Copenhagen, Copenhagen.

Clements, J. (2013), *The Value of Climate Services Across Economic and Public Sectors*.

Connor, J. et al. (2009), *Impacts of Climate Change on Lower Murray Irrigation*, Australian Journal of Agricultural and Resource Economics, Vol. 53(3), *http://dx.doi.org/10.1111/j.1467-8489.2009.00460.x*.

Considine, T.J. et al. (2004), *The value of hurricane forecasts to oil and gas producers in the Gulf of Mexico*, Journal of Applied Meteorology, 43, 1270-1281.

Crowe K.A. and W.H.Parker (2008), *Using portfolio theory to guide reforestation and restoration under climate change scenarios*, Climatic Change, Vol. 89, *http://dx.doi.org/10.1007/s10584-007-9373-x*.

Darch, G., B. Arkell and J. Tradewell (2011), *Water resource planning under climate uncertainty in London*, Atkins Report 5103993/73/DG/035 for the Adaptation Sub-Committee and Thames Water, Atkins, Epsom.

De Bel, M., A.H.H.M. Schomaker and F.C.J. van Herpen (2011), *Meerwaarde Levende Waterbouw: een maatschappelijke kostprijsanalyse*, Royal Haskoning in opdracht van Rijkswaterstaat Waterdienst.

de Bruin, K. (2014), Annex A: Details on the Adaptation Funding Gap. In. UNEP (United Nations Environment Programme) (2014), *The Adaptation Gap Report 2014*, *www.unep.org/climatechange/adaptation/gapreport2014/* (accessed 16 March 2015).

de Bruin, K. et al. (2014), *Costs and benefits of adapting spatial planning to climate change: Lessons learned from a large-scale urban development project in the Netherlands*, Regional Environmental Change, Vol. 14(3), *http://dx.doi.org/10.1007/s10113-013-0447-1*.

de Bruin, K, (2012), *Investment in flood protection measures under climate change uncertainty*, Working Paper 2012:01, CICERO, Oslo, Norge, 51pp.

de Bruin, K., R. Dellink and S. Agrawala (2009), *Economic Aspects of Adaptation to Climate Change: Integrated Assessment Modelling of Adaptation Costs and Benefits*, OECD Environment Working Papers, No. 6, OECD Publishing, Paris, *http://dx.doi.org/10.1787/225282538105*.

de Bruin, K. et al. (2009b), *Adapting to climate change in The Netherlands: An inventory of climate adaptation options and ranking of alternatives*, Climatic Change, Vol. 95(1-2), *http://dx.doi.org/10.1007/s10584-009-9576-4*.

Desjarlais, C. and C. Larrivée (2011), *Analyse économique de l'adaptation aux changements climatiques en matière de drainage urbain au Québec : comparaison de diverses stratégies d'adaptation pour un secteur de Montréal*, Consortium sur la climatologie régionale et l'adaptation aux changements climatiques, Canada.

Defra (2011), *The UK Climate Change Risk Assessment (CCRA)*, UK Department for Environment, Food and Rural Affairs, London.

Dellink, R. et al. (2014), *Consequences of Climate Change Damages for Economic Growth: A Dynamic Quantitative Assessment*, OECD Economics Department Working Papers, No. 1135, OECD Publishing, Paris, *http://dx.doi.org/10.1787/5jz2bxb8kmf3-en*.

Deltacommissie (2008), *Working together with water. A living land builds for its future*, Findings of the Deltacommissie, DeltaCommissie, The Hague.

Delta Programme (2014), *Working on the delta: Promising solutions for tasking and ambitions*, The Delta Programme, The Hague.

Delta Programme (2011), *Working on the delta: Investing in a safe and attractive Netherlands, now and in the future*, The Delta Programme, The Hague.

Desbartes, J. (2012), *Adaptation Cost Assessment Of: Early Warning Systems And Implementing Measures*, Report as part of the EC Project *Methodologies for climate proofing investments and measures under cohesion and regional policy and the common agricultural policy*, IEEP.

Desjeux, G., L. Galoisy-Guibal and C. Colin (2005), *Cost-benefit analysis of vaccination against tick-borne encephalitis among French troops*, Pharmacoeconomics, Vol. 23(9).

Dhakal, A. and A. Dixit (2013), *Economics of climate change in the water sector in Nepal. A stakeholder-focused approach. A case study of the Rupa Watershed, Kaski, Nepal. Project*: Economics of Climate Change in the Water Sector project, International Institute for Environment and Development (IIED), London, UK.

Di Falco, S. and M. Veronesi (2012), *How African Agriculture Can Adapt to Climate Change? A Counterfactual Analysis from Ethiopia*, Working Paper Series, No. 14, Department of Economics, University of Verona.

DFID (2014), *Early Value-for-Money Adaptation: Delivering VfM Adaptation using Iterative Frameworks and Low-Regret Options*, UK Department for International Development, London.

Dore, M.H.I. and I. Burton (2001), *The Costs of Adaptation to Climate Change in Canada: A Stratified Estimate by Sectors and Regions*, a report prepared for the Climate Change Action Fund, Natural Resources Canada, Ottawa.

Downing, T.E. (2012), *Views of the frontiers in climate change adaptation economics*, Wiley Interdisciplinary Reviews: Climate Change, Vol. 3(2), *http://dx.doi.org/10.1002/wcc.157*.

Downing, T. et al. (2011), *Planning and costing agriculture's adaptation to climate change: Policy Perspectives*, International Institute for Environment and Development (IIED), London, UK

EA (2011), *Thames Estuary 2100: Strategic Outline Programme*, Environment Agency (UK), Bristol.

EA (2009a), *Thames Estuary 2100: Managing flood risk through London and the Thames estuary*, TE2100 Plan, Consultation Document, Environment Agency, Bristol.

EA (2009b), *Investing for the Future Flood and Coastal Risk Management in England: A Long-term Investment Strategy*, Environment Agency, Bristol.

EA (2009c), *Using science to create a better place: Ecosystem services case studies*, Environment Agency, Bristol.

EASPE (2002), *Use and Benefits of the National Weather Service River and Flood Forecasts*, National Hydrologic Warning Council, Denver.

Ebi, K.L. (2008), *Adaptation costs for climate change-related cases of diarrhoeal disease, malnutrition, and malaria in 2030*, Global Health, Vol. 4(9), *http://dx.doi.org/10.1186/1744-8603-4-9*.

Ebi, K.L. et al. (2004), *Heat Watch/Warning Systems Save Lives: Estimated Cost and Benefits for Philadelphia 1995-98*, Bulletin of the American Meteorological Society, *http://dx.doi.org/10.1175/BAMS-85-8-1067*.

ECA (2009), *Shaping climate-resilient development: A framework for decision-making*, a report of the Economics of Climate Adaptation working group, Economics of Climate Adaptation.

ECLAC (2011a), *Review of the Economics of Climate Change in the Caribbean (RECC)*. Economic Commission for Latin America and the Caribbean (ECLAC). *http://www.cepal.org/cgi-bin/getprod.asp?xml=/portofspain/noticias/paginas/4/34444/P34444.xml&xsl=/portofspain/tpl-i/p18f.xsl&base=/portofspain/tpl-i/top-bottom.xsl*.

ECLAC (2011b), *Assessment of the Economic Impact of Climate Change on the Coastal and Human Settlements Sector In Guyana*. Economic Commission for Latin America and the Caribbean (ECLAC).

ECLAC (2011c), *An assessment of the Economic Impact of Climate Change on the Health Sector in Saint Lucia*. Economic Commission for Latin America and the Caribbean (ECLAC).

ECONADAPT (2015), *The Costs and Benefits of Adaptation*, results from the ECONADAPT Project, ECONADAPT consortium, *http://econadapt.eu/*.

EEA (2007), *Climate change: The cost of inaction and the cost of adaptation*, EEA Technical Report, No. 13/2007, European Environment Agency, Copenhagen.

Eijgenraam, C., et al (2014). *Economically Efficient Standards to Protect the Netherlands Against Flooding*, Interfaces. Vol. 44(1), *http://dx.doi.org/10.1287/inte.2013.0721*.

Epstein, P. and E. Mills (2005), *Climate Change Futures: Health, Ecological and Economic Dimensions*, The Center for Health and Global Environment, Harvard Medical School.

Environment Canada and Natural Resources Canada (2006), *Impacts of Sea Level Rise and Climate Change on the Coastal Zone of Southeastern New Brunswick*.

Evans, E. et al. (2004), *Foresight: Future Flooding*, Office of Science and Technology, London.

Faust, A.K., C. Gonseth and M. Vielle (2012), *The economic impact of climate driven changes in water availability in Switzerland*, EciMod, No. 4277.

FAO (2015), *Climate change and food systems: global assessments and implications for food security and trade*, Elbehri, A. (ed.), Food and Agriculture Organisation, Rome.

FAO (2013), *Climate Smart Agriculture*, *www.climatesmartagriculture.org*, Food and Agriculture Organisation (accessed 28 April 2015).

FDRE (2015), *Ethiopia's Climate-Resilience Strategy Agriculture*, Federal Democratic Republic of Ethiopia, Environmental Protection Authority.

Frontier Economics (2013), *The Economics of Climate Resilience CA0401*, Frontier Economics Ltd, Ecofys UK Ltd, Irbaris LLP, *www.frontier-economics.com/documents/2014/03/frontier-report-the-economics-of-climate-resilience-2.pdf* (Accessed 16 March 2015).

Fuss, S. et al. (2011), *Large-Scale Modelling of Global Food Security and Adaptation under Crop Yield Uncertainty*, Paper prepared for 2011 International Congress, August 30-September 2, 2011, Zurich, Switzerland, by European Association of Agricultural Economists, *http://purl.umn.edu/114347*.

Fuss, S. et al. (2015), *Global food security & adaptation under crop yield volatility*, Technological Forecasting and Social Change. *doi:10.1016/j.techfore.2015.03.019*.

Füssel H.M. and R.J.T. Klein (2006), *Climate Change Vulnerability Assessments: An Evolution of Conceptual Thinking*, Climatic Change, 75: 301-329.

GCAP (Global Climate Adaptation Partnership) (2011), *The Economics of Climate Change in the United Republic of Tanzania*, a report prepared for the Development Partners Group and DFID.

Gersonius, B. et al. (2013), *Climate change uncertainty: Building flexibility into water and flood risk infrastructure*, Climatic Change, Vol. 116(2), *http://dx.doi.org/10.1007/s10584-012-0494-5*.

Government of Tanzania (2014), *Agriculture Climate Resilience Plan (2014-2019)*, The United Republic of Tanzania, Ministry of Agriculture, Food Security and Cooperatives, Dar Es Salaam.

Groves, D.G. and C. Sharon (2013), *Planning Tool to Support Planning the Future of Coastal Louisiana*, Journal of Coastal Research, Special Issue 67, *http://dx.doi.org/10.2112/SI_67_10*.

Groves, D.G. et al. (2013), *Adapting to a Changing Colorado River. Making Future Water Deliveries More Reliable Through Robust Management Strategies*, Prepared for the US Bureau of Reclamation and conducted in the Environment, Energy and Economic Development Program within RAND Justice, Infrastructure and Environment, Santa Monica.

Hallegatte, S. et al. (2013), *Future flood losses in major coastal cities*, Nature Climate Change, Vol. 3, *http://dx.doi.org/10.1038/nclimate1979.*

Hamilton, J.M., D. Maddison and R.S.J. Tol (2005), *Climate change and international tourism: A simulation study, Glob Environ Chang* 15(3):253-266.

Hawley, K., M. Moench and L. Sabbag (2012), *Understanding the economics of flood risk reduction: A preliminary analysis*, Institute for Social and Environmental Transition-International; Boulder CO.

Hinkel, J. et al. (2014), *Coastal flood damage and adaptation costs under 21st century sea-level rise*, PNAS, Vol. 111(9), *http://dx.doi.org/10.1073/pnas.1222469111.*

Hinkel, J. et al. (2013), *A global analysis of erosion of sandy beaches and sea-level rise: An application of DIVA*, Global Planetary Change, Vol. 111, *http://dx.doi.org/10.1016/j.gloplacha.2013.09.002.*

Hinkel, J. et al. (2010), *Assessing risk of and adaptation to sea-level rise in the European Union: An application of DIVA*, Mitigation and Adaptation Strategies for Global Change, Vol. 15, *http://dx.doi.org/10.1007/s11027-010-9237-y.*

Hinkel, J. and R.J.T. Klein (2009), *Integrating knowledge to assess coastal vulnerability to sea-level rise: The development of the DIVA tool*, Global Environmental Change, 19 (3): 384-395, *http://dx.doi.org/10.1016/j.gloenvcha.2009.03.002.*

HKV (HKV Consultants) and RPA (Risk Policy Analysts) (2014), *Study on Economic and Social Benefits of Environmental Protection and Resource Efficiency Related to the European Semester*, a report prepared for DG Environment.

HM Government (2013), *The National Adaptation Programme Report: Analytical Annex*, Economics of the National Adaptation Programme, HM Government, London.

Hochrainer-Stigler, S. et al. (2011), *The Costs and Benefits of Reducing Risk from Natural Hazards to Residential Structures in Developing Countries*, Working Paper, No. 2011-01, The Wharton School, University of Pennsylvania, Philadelphia.

Howden, M. et al. (2003), *An overview of the adaptive capacity of the Australian agricultural sector to climate change: options, costs and benefits*, Technical Report, CSIRO Sustainable Ecosystems, Canberra.

Hsai, E.C. et al. (2002), *Cost-effectiveness of the Lyme disease vaccine*, Arthritis Rheum, Vol. 46(6), *http://dx.doi.org/10.1002/art.10270.*

Hughes, G., P. Chinowsky and K. Strzepek (2010), *The costs of adaptation to climate change for water infrastructure in OECD countries*, Utilities Policy, Vol. 18(3), *http://dx.doi.org/10.1016/j.jup.2010.03.002.*

Hunt (2011), *Policy Interventions to Address Health Impacts Associated with Air Pollution, Unsafe Water Supply and Sanitation and Hazardous Chemicals*, Environment Working Paper No. 35, OECD.

Hunt, A. and L. Anneboina, L. (2011), *Quantification and Monetisation of the Costs of Windstorms*, Deliverable 2D, ClimateCost project.

Hunt, A. and P. Watkiss (2010), *The Costs of Adaptation: Case Study on Heat Related Health Risks and Adaptation in London, Report under Specific Agreement under the Framework contract for economic support* (EEA/IEA/09/002) – Lot 2 "Climate change adaptation: costs and benefits", Report to the European Environment Agency.

IBD and ECLAC (2014), *La economía del cambio climático en el Perú*, Inter-American Development Bank, and Economic Commission for Latin America and the Carribean Lima, *http://www.cepal.org/es/publicaciones/37419-la-economia-del-cambio-climatico-en-el-peru* (accessed 20 April 2015).

ICF (2007), *The potential costs of climate change adaptation for the water industry*, Report submitted to UK Environment Agency, ICF International, London, UK.

IDS-Nepal (Integrated Development Society Nepal), PAC (Practical Action Consulting) and GCAP (Global Climate Adaptation Partnership) (2014), *Economic Impact Assessment of Climate Change in Key Sectors in Nepal*, Integrated Development Society Nepal , Kathmandu.

IFC (2011), *Climate Risk and Business: Ports – Terminal Marítimo Muelles el Bosque Cartagena, Colombia*, International Finance Corporation, Washington, DC.

Iglesias, A. et al. (2012), *Agriculture. The Climate Cost Project*, Final Report.

Ignaciuk, A. and D. Mason-D'Croz (2014), *Modelling Adaptation to Climate Change in Agriculture*, OECD Food, Agriculture and Fisheries Papers, No. 70, OECD Publishing, Paris, *http://dx.doi.org/10.1787/5jxrclljnbxq-en*.

IPCC (Intergovernmental Panel on Climate Change) (2014), *Climate Change 2014: Impacts, Adaptation, and Vulnerability*, Contribution of Working Group II to the Fifth Assessment Report of the Intergovernmental Panel on Climate Change, Cambridge University Press, Cambridge and New York.

IPCC (2013), *Climate Change 2013: The Physical Science Basis*, Contribution of Working Group I to the Fifth Assessment Report of the Intergovernmental Panel on Climate Change, Cambridge University Press, Cambridge and New York.

IPCC (2012), *Managing the Risks of Extreme Events and Disasters to Advance Climate Change Adaptation*, A Special Report of Working Groups I and II of the Intergovernmental Panel on Climate Change, Cambridge University Press, Cambridge and New York.

IPCC (2007), *Climate Change 2007: Impacts, Adaptation and Vulnerability*, Contribution of Working Group II to the Fourth Assessment Report of the Intergovernmental Panel on Climate Change, Cambridge University Press, Cambridge and New York.

Isaac, M. and D.P. van Vuuren (2009), *Modeling global residential sector energy demand for heating and air conditioning in the context of climate change*, Energy Policy, Vol. 37(2), *http://dx.doi.org/10.1016/j.enpol.2008.09.051*.

Jeuland, M. and D. Whittington (2013), *Water Resources Planning under Climate Change: A 'Real Options' Application to Investment Planning in the Blue Nile*, Environment for Development. Discussion Paper Series, No. 13-5.

Jochem, E. and W. Schade (2009), *Adaptation and Mitigation Strategies Supporting European Climate Policy*, ADAM Deliverable D-M1.2, Fraunhofer Institute for Systems and Innovation Research, Munich.

Jongman, B. et al. (2014), *Increasing Stress on Disaster-Risk Finance due to Large Floods*, Nature Climate Change, Vol. 4, *http://dx.doi.org/10.1038/nclimate2124*.

Jonkeren, O. (2009), *Adaption to climate change in inland waterway transport*, Vrije Universiteit.

Kettunen, M. (2011), *Water, ecosystem services and nature: putting the 'green' into green economy*, Paper presented at the Stockholm World Water Week, 23 August 2011, Stockholm.

Khabarov, N. et al. (2014), *Forest fires and adaptation options in Europe*, Regional Environmental Change, *http://dx.doi.org/10.1007/s10113-014-0621-0*.

Kind, J. M (2014). *Economically efficient flood protection standards for the Netherlands*. Flood Risk Management, 7 (2014) 103-117. *DOI: 10.1111/jfr3.12026*.

Kirshen, P. (2007), *Adaptation Options and Cost in Water Supply*, Report to the UNFCCC Secretariat, Bonn.

Kirshen, P. et al. (2005), *Global Analysis of Changes in Water Supply Yields and Costs under Climate Change: A Case Study in China*, Climatic Change, Vol. 68(3), *http://dx.doi.org/10.1007/s10584-005-1148-7*.

Kirshen. P. et al. (2004), *Climate's Long-term Impacts on Metro Boston (CLIMB)*, Final Report to the US EPA. Available at *http://dx.doi.org/10.1007/s10584-005-1148-7*.

KMatrix (2014), *London's Adaptation Economy 2012/13*, Report to the Mayor of London.

Kontogianni, A. et al. (2013), *Assessing sea level rise costs and adaptation benefits under uncertainty in Greece*, Environmental Science & Policy, vol. 37, *http://dx.doi.org/10.1016/j.envsci.2013.08.006*.

Kull, D., R. Mechler and S. Hochrainer-Stigler (2013), *Probabilistic cost-benefit analysis of disaster risk management in a development context*, Disasters, Vol. 37(3), *http://dx.doi.org/10.1111/disa.12002*.

Kundzewicz, Z.W. et al. (2014), *Flood risk and climate change: Global and regional perspectives*, Hydrological Sciences Journal, Vol. 59(1), *http://dx.doi.org/10.1080/02626667.2013.857411*.

Lazo, J.K., J.S. Rice and M.L. Hagenstad (2010), *Benefits of investing in weather forecasting research: An application to supercomputing*, Yuejiang Academic Journal, Vol. 2.

Lazo, J.K. and D.M. Waldman (2011), *Valuing improved hurricane forecasts*, Economics Letters, Vol. 111(1), *http://dx.doi.org/10.1016/j.econlet.2010.12.012*.

LCCP (2009), *Economic Incentive Schemes for Retrofitting London's Existing Homes for Climate Change Impacts*, London Climate Change Partnership, Published by Greater London Authority.

Leclère, D. et al. (2014), *Climate change induced transformations of agricultural systems: insights from a global model*, Environmental Research.Letters, Vol. 9(12), *http://dx.doi.org/10.1088/1748-9326/9/12/124018*.

Lempert, R.J. and D.G. Groves (2010), *Identifying and evaluating robust adaptive policy responses to climate change for water management agencies in the American west*, Technological Forecasting and Social Change, Vol. 77(6), *http://dx.doi.org/10.1016/j.techfore.2010.04.007*.

Lempert, R.J. et al. (2013), *Ensuring Robust Flood Risk Management in Ho Chi Minh City*, Policy Research Working Papers, No. 6465, World Bank, Washington, DC, *http://dx.doi.org/10.1596/1813-9450-6465*.

Liao, K.J. et al. (2010), *Cost Analysis of Impacts of Climate Change on Regional Air Quality*, Journal of the Air & Waste Management Association, Vol. 60(2), *http://dx.doi.org/10.3155/1047-3289.60.2.195*.

Lunduka, R.W. (2013), *Multiple stakeholders' economic analysis of climate change adaptation: A case study of Lake Chilwa Catchment, Malawi*, IIED, London.

Macauley, M.K. (2010), *Climate Adaptation Policy: The Role and Value of Information*, Issue Brief, No. 10-10, Resources for the Future, Washington, DC.

Máñez, M. and A. Cerdà (2014), *Prioritisation Method for Adaptation Measures to Climate Change in the Water Sector*, CSC Report 18, Climate Service Center, Germany.

Margulis, S. and C.B.S. Dubeux (2010), *Economia da Mudança do Clima no Brasil: Custos e Oportunidades*, IBEP Gráfica, São Paulo.

Markandya, A., I. Galarraga and E.S. de Murieta (eds.) (2014), *Routledge Handbook of the Economics of Climate Change Adaptation*, Routledge, Oxon and New York.

Markandya, A. and A. Mishra (eds.) (2010), *Costing Adaptation: Preparing for climate change in India*, The Energy and Resources Institute, New Delhi.

Martin-Ortega, J. et al. (2012), *Cost-effectiveness analysis report for the Thame sub-catchment including analysis of disproportionality*. FP7 REFRESH Adaptive strategies to Mitigate the Impacts of Climate Change on European Freshwater Ecosystems. Deliverable 6.12.

Matiya, G., R. Lunduka and M. Sikwese (2011), *Planning and costing agricultural adaptation to climate change in the small-scale maize production system of Malawi*, International Institute for Environment and Development (IIED), London, UK, *http://pubs.iied.org/G03170.html* (Accessed 16 March 2015).

McCarl, B.A. (2007), *Adaptation Options for Agriculture, Forestry and Fisheries*, a report to the UNFCCC Secretariat, Financial and Technical Support Division.

McCarthy, N., L. Lipper and G. Branca (2011), *Climate-Smart Agriculture: Smallholder Adoption and Implications for Climate Change Adaptation and Mitigation*, Mitigation of Climate Change in Agriculture Series, No. 4, FAO, Rome.

Mechler, R. (2012), *Reviewing the economic efficiency of disaster risk management*, Review Commissioned by Foresight Project: Reducing Risks of Future Disasters.Priorities for Decision Makers, IIASA, 2012.

Mechler, R. (2009), *The Cost-Benefit Analysis Methodology*, From Risk to Resilience Working Paper, No. 1, Moench, M., E. Caspari and A. Pokhrel (eds.), ISET, ISET-Nepal and ProVention, Kathmandu.

Mechler, R. (2005), *Cost-benefit analysis of natural disaster risk management in developing countries*, Deutsche Gesellschaft für Technische Zusammenarbeit (GTZ). Arbeitskonzept, Eschborn, Germany.

Mechler, R. et al. (2014), *Making Communities More Flood Resilient: The Role of Cost Benefit Analysis and Other Decision-Support Tools in Disaster Risk Reduction*, White Paper, Zurich Flood Resilience Alliance.

Metroeconomica (2006), *Climate Change Impacts and Adaptation: Quantify the Cost of Impacts and Adaptation*, Report to Defra, London.

Michelon, T., P. Magne and F. Simon-Delaville (2005), *Lessons of the 2003 heat wave in France and action taken to limit the effects of future heat waves*, in Kirch, W., B. Menne and R. Bertollini (eds.), *Extreme Weather Events and Public Health Responses*, Springer-Verlag.

Mima, S., P. Criqui and P. Watkiss (2011), *The Impacts and Economic Costs of Climate Change on Enery in Europe: Summary of Results from the EC RTD ClimateCost Project*, in Watkiss, P. (ed.), *The ClimateCost Project*, Vol. 1, Stockholm Environment Institute, Sweden.

MMC (2005), *Natural Hazard Mitigation Saves: An Independent Study to Assess the Future Savings from Mitigation Activities*, Volume 2 – Study Documentation, Multihazard Mitigation Council, Washington, DC.

Moench, M. et al. (2009), *Rethinking the costs and benefits of disaster risk reduction under changing climate conditions*, in Moench, M. et al. (eds.), *Catalyzing Climate and Disaster Resilience*, ISET-Nepal, Kathmandu.

Moran, D. et al. (2013), *Research to assess preparedness of England's natural resources for a changing climate: Part 2 – assessing the type and level of adaptation action required to address climate risks in the 'vulnerability hotspots'*, Report to the Adaptation Sub-Committee of the Committee on Climate Change, United Kingdom.

Mosnier, A. et al. (2014), *Global food markets, trade and the cost of climate change adaptation*, Food Security, Vol. 6(1), *http://dx.doi.org/10.1007/s12571-013-0319-z*.

Muller, M. (2007), *Adapting to Climate Change: Water Management for Urban Resilience*, Environment and Urbanization, Vol. 19(1), *http://dx.doi.org/10.1177/0956247807076726*.

Nassopoulos, H., P. Dumas and S. Hallegatte (2013), *Adaptation to an Uncertain Climate Change: Cost Benefit Analysis and Robust Decision Making for Dam Dimensioning*, Climatic Change (2012) 114:497-508, *http://dx.doi.org/10.1007/s10584-012-0423-7*.

Naumann, S. et al. (2011), *Assessment of the potential of ecosystem-based approaches to climate change adaptation and mitigation in Europe*, Final report to the European Commission, DG Environment, Ecologic Institute, Berlin, and the Environmental Change Institute, Oxford University Centre for the Environment, Oxford.

Neufeldt, H. et al., (2010), *Climate policy and inter-linkages between adaptation and mitigation*, Chapter 1. In. *Making Climate Change Work for Us. European perspectives of adaptation and mitigation strategies* (ADAM). Editors: Mike Hulme and Henry Neufeldt. Published by Cambridge University Press, 2010.

Neumann, J. et al. (2011), *The economics of adaptation along developed coastlines*, Wiley Interdisciplinary Reviews: Climate Change, Vol. 2(1), *http://dx.doi.org/10.1002/wcc.90*.

Nelson, G.C. et al. (2009), *Climate Change: Impact on Agriculture and Costs of Adaptation*, IFPRI, Washington, DC.

NRC (Natural Resources Canada) (2007), *From Impacts to Adaptation: Canada in a Changing Climate*, Government of Canada, Ottawa.

NRTEE (2011), *Paying the Price: The Economic Impacts of Climate Change for Canada*, National Round Table on the Environment and the Economy, Ontario.

Nurmi, V. et al. (2013), *Cost-Benefit Analysis Of Green Roofs In Urban Areas: Case Study In Helsinki*. Finnish Meteorological Institute. Helsinki 2013. Available at *https://helda.helsinki.fi/bitstream/handle/10138/40150/2013nro2.pdf?sequence=1* (Accessed March 2015).

OECD (2007), *Climate Change in the European Alps: Adapting Winter Tourism and Natural Hazards Management*, OECD Publishing, Paris, *http://dx.doi.org/10.1787/9789264031692-en*.

OECD (2008), *Economic Aspects of Adaptation to Climate Change: Costs, Benefits and Policy Instruments*, OECD Publishing, Paris, *http://dx.doi.org/10.1787/9789264046214-en*.

ONERC (2009), *Climate change: costs of impacts and lines of adaptation*, report to the Prime Minister and Parliament, Observatoire National sur les Effets du Réchauffement Climatique, Paris.

Ostro, B. et al. (2010), *The Effects of Temperature and Use of Air Conditioning on Hospitalizations*, American Journal of Epidemiology, 2010, Vol. 172, No. 9, *http://dx.doi.org/10.1093/aje/kwq231*.

Parry, M. et al. (2009), *Assessing the costs of Adaptation to climate change: A review of the UNFCCC and other recent estimates*, International Institute for Environment and Development and Grantham Institute for Climate Change, London.

Parry, M. et al. (2004), *Effects of climate change on global food production under SRES emissions and socio-economic scenarios*, Global Environmental Change, Vol. 14(1), *http://dx.doi.org/10.1016/j.gloenvcha.2003.10.008*.

Paul, B.K. (2009), *Why Relatively Fewer People Died? The Case of Bangladesh's Cyclone Sidr*, Natural Hazards, Vol. 50, *http://dx.doi.org/10.1007/s11069-008-9340-5*.

Pilli-Sihlova, K. et al. (2010), *Climate change and electricity consumption – Witnessing increasing or decreasing use and costs?*, Energy Policy, Vol. 38(5), *http://dx.doi.org/10.1016/j.enpol.2009.12.033*.

Pohl, I. et al. (2014), *CBA Rotterdam Climate Adaptation Strategy – Case: Bergpolder Zuid*, consultancy report by REBEL to City of Rotterdam, Office for Sustainability and Climate Change. Rotterdam.

Porter, et al. (2014), *Food security and food production systems*. In: *Climate Change 2014: Impacts, Adaptation, and Vulnerability. Part A: Global and Sectoral Aspects*. Contribution of Working Group II to the Fifth Assessment Report of the Intergovernmental Panel on Climate Change. Cambridge University Press, Cambridge, United Kingdom and New York, NY, USA, pp. 485-533.

Price, D. and K. Isaac (2012), *Adapting Sustainable Forest Management to Climate Change: Scenarios for Vulnerability Assessment*, Canadian Council of Forest Ministers, Ottawa.

Ranger, N. and S.-L. Garbett-Shiels (2011), *How can decision-makers in developing countries incorporate uncertainty about future climate risks into existing planning and policy-making processes?*, Policy Paper, Grantham Research Institute on Climate Change and the Environment, London.

Ranger, N. et al. (2011), *An assessment of the potential impact of climate change on flood risk in Mumbai*, Climatic Change, Vol. 104, *http://dx.doi.org/10.1007/s10584-010-9979-2*.

RH DHV (2012), *Costs and Benefits of Sustainable Drainage Systems*, Report to the CCC by Royal Haskoning DHV.

Rojas, R., L. Feyen and P. Watkiss (2013), *Climate change and river floods in the European Union: Socio-economic consequences and the costs and benefits of adaptation*, Global Environmental Change, Vol. 23(6), *http://dx.doi.org/10.1016/j.gloenvcha.2013.08.006*.

RMS (Risk Management Solutions) (2009), *Analyzing the Effects of the My Safe Florida Home Program on Florida Insurance Risk: RMS Special Report*, Risk Management Solutions Inc.

Scandizzo, P.L. (2011), *Climate Change Adaptation and Real Option Evaluation*, CEIS Working Paper, No. 232, *http://dx.doi.org/10.2139/ssrn.2046955*.

SCCV (Swedish Commission on Climate and Vulnerability) (2007), *Sweden facing climate change - threats and opportunities*, Ministry of the Environment and Energy, Stockholm.

Seo, S.N. et al. (2009), *A Ricardian Analysis of the Distribution of Climate Change Impacts on Agriculture across Agro-Ecological Zones in Africa*, Environmental and Resource Economics, Vol. 43(3), *http://dx.doi.org/10.1007/s10640-009-9270-z*.

SEI (Stockholm Environment Institute) (2009), *The Economics of Climate Change in East Africa*: Downing, T. et al. Report for DFID and DANIDA. Available at *http://kenya.cceconomics.org/*.

Settele, J. et al. (2014), *Terrestrial and inland water systems*. In: *Climate Change 2014: Impacts, Adaptation, and Vulnerability. Part A: Global and Sectoral Aspects*. Contribution of Working Group II to the Fifth Assessment Report of the Intergovernmental Panel on Climate. Cambridge University Press, Cambridge, United Kingdom and New York, NY, USA, pp. 271-359.

Sgobbi, A. and C. Carraro (2008), *Climate Change Impacts and Adaptation Strategies in Italy: An Economic Assessment*, FEEM Fondazione Eni Enrico Mattei Research Paper; CMCC Research Paper, No. 14, *http://dx.doi.org/10.2139/ssrn.1086627*.

Skourtos, M. et al. (2014), *Incorporating cross-sectoral effects into analysis of the cost-effectiveness of climate change adaptation measures*, Climatic Change, Vol. 128(3-4), *http://dx.doi.org/10.1007/s10584-014-1168-2*.

Skourtos, M., A. Kontogianni and C. Tourkolias (2013), *Report on the Estimated Cost of Adaptation Options Under Climate Uncertainty*, The CLIMSAVE Project – Climate Change Integrated Assessment Methodology for Cross-Sectoral Adaptation and Vulnerability in Europe, University of the Aegean, Greece.

Smith, et al. (2014), *Human health: Impacts, adaptation, and co-benefits*. In: *Climate Change 2014: Impacts, Adaptation, and Vulnerability. Part A: Global and Sectoral Aspects*. Contribution of Working Group II to the Fifth Assessment Report of the Intergovernmental Panel on Climate Change. Cambridge University Press, Cambridge, United Kingdom and New York, NY, USA, pp. 709-754.

Stanton, E.A., M. Davis and A. Fencl (2010), *Costing Climate Impacts and Adaptation: A Canadian Study on Coastal Zones*, a report commission by the National Round Table on the Environment and the Economy, Stockholm Environment Institute, Somerville.

Stern, N. et al. (2006), *The Economics of Climate Change*, HM Treasury, London.

Subbiah, A., L. Bildan and R. Narasimhan (2008), *Background Paper on Assessment of the Economics of Early Warning Systems for Disaster Risk Reduction*, submitted to the World Bank Group, Global Facility for Disaster Reduction and Recovery, World Bank, Washington, DC.

Sussman, F. et al. (2014), *Climate change adaptation cost in the US: What do we know?*, Climate Policy, Vol. 14(2), *http://dx.doi.org/10.1080/14693062.2013.777604*.

Tanaka, S.K. et al. (2006), *Climate Warming and Water Management Adaptation for California*, Climatic Change, Vol. 76(3-4), *http://dx.doi.org/10.1007/s10584-006-9079-5*.

Tainio, A. et al. (2014), *Conservation of grassland butterflies in Finland under a changing climate*, Regional Environmental Change, *http://dx.doi.org/10.1007/s10113-014-0684-y*.

TEEB (The Economics of Ecosystems & Biodiversity) (2010), *The Economics of Ecosystems and Biodiversity for Local and Regional Policy Makers*, Progress Press, Malta.

TEEB (2009), *TEEB Climate Issues Update*, UNEP, Nairobi, *www.teebweb.org/media/2009/09/TEEB-Climate-Issues-Update.pdf* (accessed 26 February 2015).

TEEB DE (2014), *Naturkapital und Klimapolitik – Synergien und Konflikte, Kurzbericht für Entscheidungsträger*, (Natural capital and climate policy - synergies and conflicts, Summary report for decision-makers), Technische Universität Berlin, HelmholtzZentrum für Umweltforschung – UFZ, Leipzig.

Teichmann, M. and A. Berghöfer (2010), *TEEBcase River Elbe flood regulation options with ecological benefits, Germany*, *www.teebweb.org* (accessed 26 February 2015).

Triple E Consulting (2014), *Assessing the Implications of Climate Change Adaptation on Employment in the EU*, Report to the European Commission, DG Climate Action.

Tröltzsch, J., H. Lückge and C. Sartorius (2012), *Kosten und Nutzen von Anpassungsmaßnahmen an den Klimawandel: Analyse von 28 Anpassungsmaßnahme in Deutschland* [Costs and benefits of climate adaptation measures. Analysis of 28 adaptation measures in Germany]. Climate Change, No. 10/2012, UBA, Dessau.

UKCIP (UK Climate Impacts Programme) (2006), *Identifying adaptation options*, UKCIP, London.

UNDP (United Nations Development Programme) (2011), *Assessment of Investment and Financial Flows to Address Climate Change* (Capacity Development for Policy Makers to Address Climate Change). Country summaries. Available at: *www.undpcc.org/en/financial-analysis/results* (accessed 3 March 2015).

UNEP (2014), *The Adaptation Gap Report 2014*, United Nations Environment Programme, *www.unep.org/climatechange/adaptation/gapreport2014/* (Accessed 16.March.2015)

UNFCCC (2010), *Synthesis report on the National Economic, Environment and Development Study (NEEDS) for Climate Change Project*, *FCCC/SBI/2010/INF.7*, United Nations Framework Convention on Climate Change, Bonn.

UNFCCC (2009), *Potential costs and benefits of adaptation options: A review of existing literature*, *FCCC/TP/2009/2*, United Nations Framework Convention on Climate Change, Bonn.

UNFCCC (2007), *Investment and Financial Flows to Address Climate Change*, United Nations Framework Convention on Climate Change, Bonn.

van Ierland, E.C. et al. (2006), *A qualitative assessment of climate change adaptation options and some estimates of adaptation costs*. Routeplanner subprojects 3, 4 and 5, Wageningen UR.

Vafeidis, A.T. et al. (2008), *A New Global Coastal Database for Impact and Vulnerability Analysis to Sea-Level Rise*, Journal of Coastal Research, Vol. 24(4), *http://dx.doi.org/10.2112/06-0725.1*.

van der Pol, T.D., E.C. van Ierland and H.P. Weikard (2013), *Optimal dike investments under uncertainty and learning about increasing water levels*, Journal of Flood Risk Management, Vol. 7(4), *http://dx.doi.org/10.1111/jfr3.12063*.

Vergara, W. et al. (2007), *Economic impacts of rapid glacier retreat in the Andes*, EOS Transactions American Geophysical Union, Vol. 88 (25), *http://dx.doi.org/10.1029/2007EO250001*.

Vivid Economics (2013), *The macroeconomics of climate change*, a report prepared for Defra, *www.vivideconomics.com/index.php/publications/the-macroeconomics-of-climate-change* (accessed 2 March 2015).

Ward, P.J. et al. (2010), *Partial costs of global climate change adaptation for the supply of raw industrial and municipal water: A methodology and application*, Environmental Research Letters, Vol. 5(4), *http://dx.doi.org/10.1088/1748-9326/5/4/044011*.

Watkiss, P. and A. Hunt (2010), *Review of adaptation costs and benefits estimates in Europe for the European Environment State and Outlook Report 2010*, Technical Report prepared for the European Environmental Agency for the European Environment State and Outlook Report 2010, published online by the European Environment Agency, Copenhagen.

Watkiss, P. and A. Hunt (2011), *Method for the UK Adaptation Economic Assessment* (Economics of Climate Resilience), *Final Report to Defra*, May 2011, Deliverable 2.2.1.

Watkiss, P. and A. Hunt (2014), *Low and no-regret adaptation – what do we know?*, ECONADAPT *http://adaptationfrontiers.eu/pdf/B_U_03_PWatkiss.pdf* (accessed 16 March 2015).

Watkiss, P. (ed.) (2012), *The ClimateCost Project*, Final Report, Volume 1: Europe, Stockholm Environment Institute, Sweden.

WHO (2013), *Climate change and health: A tool to estimate health and adaptation costs*, World Health Organisation Regional Office for Europe, Bonn.

WHO (2009), *Improving public health responses to extreme weather/heat-waves – EuroHEAT*, Technical Summary, World Health Organisation Regional Office for Europe, Copenhagen.

Wilby, R.L and R. Keenan (2012), *Adapting to flood risk under climate change*, Progress in Physical Geography, Vol. 36(3), *http://dx.doi.org/10.1177/0309133312438908*.

World Bank (2013), *Weather, Climate and Water Hazards and Climate Resilience: Effective Preparedness through National Meteorological and Hydrological Services*, World Bank, Washington, DC.

World Bank (2012), *A Cost Effective Solution to Reduce Disaster Losses in Developing Countries: Hydro-Meteorological Services, Early Warning, and Evacuation*, World Bank, *Policy Research Working Paper* 6058, May 2012, World Bank, Washington, DC.

World Bank (2011), *Natural Hazards, UnNatural Disasters: The Economics of Effective Prevention*, World Bank, Washington, DC.

World Bank (2010), *Economics of Adaptation to Climate Change – Ecosystem Services*, Discussion Paper, No. 7, World Bank, Washington, DC.

World Bank (2010), *The Costs to Developing Countries of Adapting to Climate Change: New Methods and Estimates*, The Global Report of the Economics of Adaptation to Climate Change Study, Synthesis Report, World Bank, Washington, DC.

World Bank (2010b), *Mozambique – Economics of adaptation to climate change*, World Bank, Washington, DC.

World Bank (2010c), *Ghana – Economics of adaptation to climate change*, World Bank, Washington, DC.

World Bank (2010d), *Samoa – Economics of adaptation to climate change*, World Bank, Washington, DC.

World Bank (2010e), *Ethiopia – Economics of adaptation to climate change*, World Bank, Washington, DC.

World Bank (2010f), *Vietnam – Economics of adaptation to climate change*, World Bank, Washington, DC.

World Bank (2010g), *Bangladesh – Economics of adaptation to climate change*, World Bank, Washington, DC.

World Bank (2009), *Evaluating Adaptation via Economic Analysis*, Guidance Note 7, Annex 12, World Bank, Washington, DC.

World Bank (2006), *Clean Energy and Development: Towards an Investment Framework*, ESSD-VP/I-VP, April 5, 2006. Washington, DC, The World Bank.

Wreford, A. and A. Renwick (2012), *Estimating the costs of climate change adaptation in the agricultural sector*, CAB Reviews, Vol. 7 (40), *http://dx.doi.org/10.1079/PAVSNNR20127040*.

Wright, L. et al.. (2012), *Estimated effects of climate change on flood vulnerability of U.S. bridges*. Mitigation and Adaptation Strategies for Global Climate Change, 17, 939-955.

Zachariadis, T. (2010), *Forecast of electricity consumption in Cyprus up to the year 2030: The potential impact of climate change*, Energy Policy, Vol. 38(2), *http://dx.doi.org/10.1016/j.enpol.2009.10.019*.

Zhou, F. et al. (2007), *Potential climate change-induced permafrost degradation and building foundations: An assessment of impacts and costs for five case communities in the Northwest Territories*.

Zhu, X. et al. (2013), *A model-based assessment of adaptation options for Chianti wine production in Tuscany (Italy) under climate change*, Regional Environmental Change, *http://dx.doi.org/10.1007/s10113-014-0622-z*.

Climate Change Risks and Adaptation
Linking Policy and Economics

Chapter 4

Framing risk-based approaches to adaptation planning

This chapter outlines a framework for applying a risk-based approach to climate change. It explores: how risks can be identified through tools including climate risk and vulnerability assessments; how those identified risks can be characterised; and how appropriate policies can be explored and chosen to address risks. It emphasises the importance of regular monitoring and evaluation in a risk-based approach, particularly given pervasive uncertainty about the future climate.

Key messages

- There are four stages to applying a risk-based approach: i) identifying relevant risks, ii) characterising those risks, iii) choosing and exploring policy options to address the risks, and iv) providing feedback to respond to evolving risks through an iterative process.
- Identifying and characterising risks require both quantitative information and qualitative judgements. The acceptability of different risks will depend upon context and value judgements, which make it essential to have a transparent and inclusive process of stakeholder engagement.
- Methodologies used to characterise risks in one sector will not necessarily be suitable for others. Tools such as systems mapping and horizon scanning can assist with understanding risks that are complex or highly uncertain.
- Feedback and learning mechanisms should be considered from the outset when applying a risk-based approach. Building in flexibility at the outset of applying a risk-based approach, and responding to changing information, can reduce costs and enhance effectiveness.

A framework for a risk-based approach to adaptation planning

The risks arising from climate change are diverse and context specific, ranging from shifting ecological zones to changes in the incidence of diseases. A flexible, transparent and iterative framework for adaptation planning can help policy makers consider a broad range of risks and evolving information about climate risks, and the responses of society, economy and ecosystems to a changing climate. This flexibility can, in turn, inform adaptation planning processes that are effective in preparing society for current climate vulnerability while accounting for factors that may increase future hazards, vulnerability and exposure.

Risk-based approaches to adaptation planning provide a systematic approach to understanding, characterising and managing climate risks. Risk-based approaches have traditionally been applied to contexts where the frequency and intensity of events are reasonably well understood (e.g. water management). When applied to the context of climate change, however, a broader and more uncertain set of risks must be considered. At the same time, climate and non-climate risks are embedded in the larger socio-economic, political and cultural context (OECD, 2014a). To better manage these multi-dimensional risks and their consequences, both systemic risks and the broader contexts of those risks must be explicitly addressed in the adaptation planning process.

This framework proposes four steps for managing climate risks: i) identifying the risks, ii) characterising the risks, iii) choosing and exploring adaptation policies to address the risks, and iv) responding to evolving risks through an iterative process of feedback and learning (see Figure 4.1). The framework proposed in this chapter builds on a three-stage framework proposed by the OECD for the management of water risks and security, to

Figure 4.1. **Four steps in a risk-based approach**

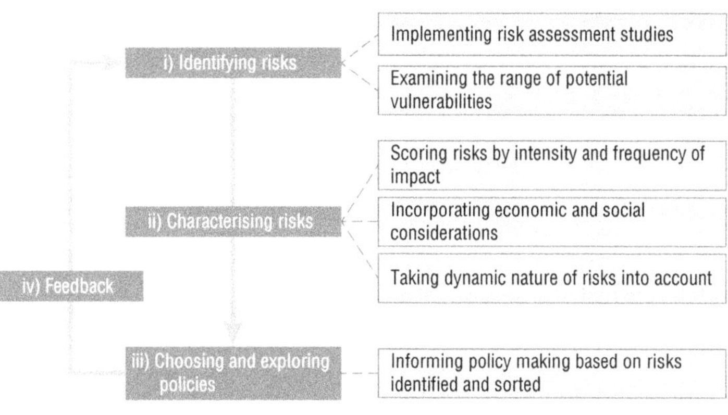

"know, target and manage" risks (OECD, 2013a). The chapter also builds on the OECD's work on risks and resilience such as "Boosting Resilience through Innovative Risk Governance" (OECD 2014a) and "Guidelines for Resilience Systems Analysis" (OECD, 2014b).

i) Identifying risks

The first stage of a risk-based approach is to identify the range of potential risks through, for example, climate risk and vulnerability assessments. The IPCC (2014) defines climate-related risks as the "potential for consequences where something of value is at stake and where the outcome is uncertain". Uncertainty in this context includes situations where the outcome is unknown, but the range of probabilities and outcomes are known: the equivalent of rolling a dice. However, uncertainty also encompasses situations where neither the full range of consequences nor their associated likelihoods can be perfectly known in advance.

The IPCC definition has several implications for the identification of risks. The first is that judgements will inevitably have to be made about which risks to include, given the wide breadth of potential impacts of climate change. In addition, the assessment process needs to incorporate judgements about the values used to assess potential risks; risks to some people can represent opportunities for others. For instance, measures to address the risk of water shortages for some communities by diverting water flows might undermine the availability of freshwater systems for other communities (OECD, 2014a). It is therefore important to be transparent and engage all relevant stakeholders in the process of identifying key risks (GIZ, 2013).

Common approaches for identifying risks include "top-down" and "bottom-up" risk assessments (Dessai and Hulme, 2004). Top-down assessments use climate change projections as the starting point for estimating potential impacts (e.g. changes in disease prevalence or crop yields). Once a list of risks has been identified, they can be translated into a common metric or filtered as part of a consultative process. Literature reviews, quantitative modelling, previous experiences of extreme weather and expert judgement can all assist in the identification of potential impacts. Bottom-up assessments, by contrast, start by analysing the characteristics of the people (or ecosystems) that will be affected by climate change. This includes an examination of the drivers of vulnerability (e.g. unsustainable water management systems or inequality) and can be complemented with an analysis of the objectives valuable to those within the study area (e.g. livelihood protection). Based on this information, bottom-up assessments can analyse how identified objectives may be affected by climate change. For both top-down and bottom-up assessments, the identification of risks involves the examination of three key factors characterising risks: hazards, exposure and vulnerability. Table 4.1 outlines a few questions that can help characterise each of these factors.

Table 4.1. **Key questions for identifying risks**

Factors	Key questions
Hazards	What are the current and projected characteristics of events (hazards) that could have significant (harmful) consequences? What are the range of potential outcomes and the estimated likelihood of them occurring? How can data on hazards be collected and disseminated?
Exposure	Who/what can be adversely affected by events caused by climate change? • *Physical*: Buildings and infrastructure • *Human and social*: The health and social fabric of the population, including physical health, health infrastructure, security, and social equity • *Economic and financial*: Properties, capital stock, sources of income, productivity, level of financial protection (e.g., insurance) and income equality • *Environmental*: Natural resources
Vulnerability	What are the characteristics of exposed populations, assets, resources and institutions? What are the factors that render populations, assets, resources and institutions vulnerable to damage? What are the estimated potential consequences from hazards, including physical, human, financial and economic, social, and environmental hazards? What measures and/or infrastructure can reduce exposure and vulnerability?

Source: Based on IPCC (2014), *Climate Change 2014: Impacts, Adaptation, and Vulnerability*, Contribution of Working Group II to the Fifth Assessment Report of the Intergovernmental Panel on Climate Change, Field, C.B. et al. (eds.), Cambridge University Press, Cambridge, United Kingdom and New York, NY, USA; OECD (2012), *OECD Environmental Outlook to 2050: The Consequences of Inaction*, OECD Publishing, Paris; Hammill, A. and T. Tanner (2011), *Harmonising Climate Risk Management: Adaptation Screening and Assessment Tools for Development Co-operation*, OECD Environment Working Papers, No. 36, OECD Publishing, Paris.

Stakeholder engagement is particularly important for the implementation of a bottom-up assessment, since the outcome of a risk and vulnerability assessment can differ depending on whom the assessment focuses on (i.e. whose values are included and whose vulnerability is analysed). To reduce the risk of such bias, it is important to engage a diverse set of stakeholders that experience vulnerability and exposure to hazards in different ways (Ayers and Forsyth, 2009).

The quality of the evidence base for climate and non-climate risks affects the resources required to undertake more detailed analysis of the risks. Data requirements include: the availability of climate change projections and the confidence that can be placed on these estimates; understanding of the responsiveness of biophysical systems; and the translation of such information into relevant metrics (e.g. monetary values). The results from the OECD country survey show that all of the respondent countries have applied analytical tools to assess the consequences of current or future climate change, or both (see Table 4.2). The survey results, however, also highlight that the main focus to date has been on qualitative analysis, with much narrower coverage of monetary and quantitative analyses of projected impacts.

Existing data on climate and non-climate risks are not always sufficiently detailed to analyse and project the diverse potential consequences of climate change. In particular, local- or regional-level risk data remain relatively sparse compared to similar data at the national level in many OECD countries (EC, 2014). Similarly, the survey of officials in OECD countries and the European Commission identified the sectors that are priorities for collecting more data (Figure 4.2). Priority areas include agriculture, water, coastal areas and

Table 4.2. **Status of the evidence base for climate-related risks at the national level**

	Identifying the impacts of current climate variability:			Projecting the impacts of climate change		
	Monetary impacts	Quantitative analysis	Qualitative analysis	Monetary impacts	Quantitative analysis	Qualitative analysis
Austria						
Belgium						
Canada						
Chile						
Czech Republic						
Denmark						
Estonia						
EU Commission						
Finland						
Greece						
Hungary						
Ireland						
Italy						
Japan						
Korea						
Netherlands						
New Zealand						
Norway						
Poland						
Portugal						
Slovak Republic						
Slovenia						
Spain						
Sweden						
Switzerland						
Turkey						
United Kingdom						

No national risk assessment has been undertaken to provide evidence base.
Some examples of risk assessment have been provided, but systematic analysis is not implemented.
Systematic analysis has been undertaken at the national level.

Figure 4.2. **Sectoral priorities for enhancing the evidence base**[1]

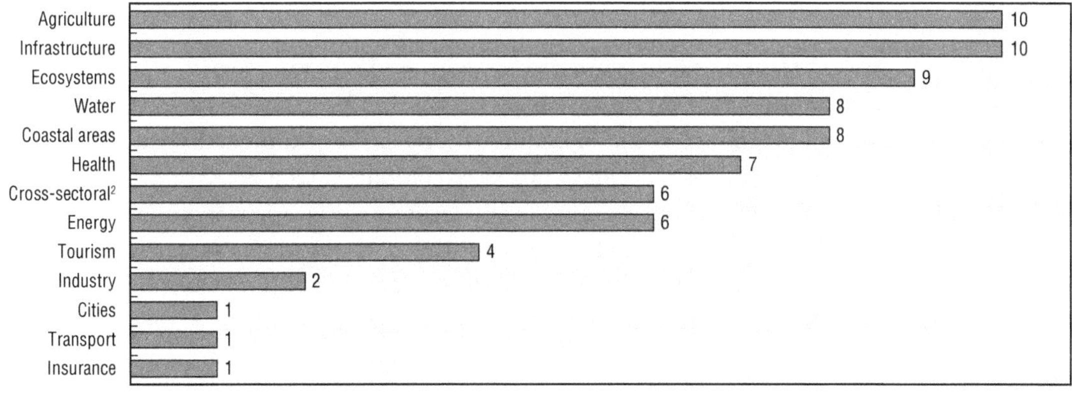

1. The number of respondents who selected that sector as a priority for enhancing the evidence base (multiple answers). 17 countries and the European Commission responded.
2. "Cross-sectoral" includes cascading effects, trade and supply chain effects.

infrastructure. Notably, the sectors identified for further research were those that already have a comparatively strong evidence base (Chapter 3).

Recognising the time and resources that may be needed to develop a more detailed evidence base, it is important not to delay an assessment of the risks until that information becomes available. Data and information already available can inform an initial assessment. At the same time, when conducting qualitative or quantitative assessment, it is useful to carefully cross-check the findings with different stakeholder groups to compare different perceptions of the risks. Further it is important to communicate what data gaps exist so that a decision can be made whether and how these gaps can be overcome (OECD, 2014b).

Assessments of climate risks should aim to consider the risks that arise from interactions within and between systems. These systemic risks can have consequences beyond the sectors and populations in the area that was directly affected by the climate impact. For example, the 2011 floods in Thailand led to 800 deaths and large economic losses domestically. In addition, these floods also disrupted the global supply chain of manufacturers, the automobile sector in particular, that relied on Thailand as a supplier (WWF and RSA, 2014). It is challenging to identify systemic risks using traditional risk assessment tools, such as literature reviews and expert judgement. Instead, it can be useful to apply risk identification measures across sectors (e.g. cross-sector workshops) or across responsible ministries (e.g. horizontal scanning). Such cross-cutting approaches can facilitate a discussion on how individual risks in one area may influence those in others, and how such risks could evolve over time as a result of these interactions.

More systematic techniques could also be applied to understand the potential impacts arising from indirect or cross-sectoral risks. "Horizon scanning" is a systematic examination of the information available to identify potential risks and to better understand how emerging trends and developments potentially may affect current policy and practice. Another approach is "systems mapping", which identifies how factors that influence present (and possibly also future) risks are linked. "Systems maps" can be used to describe the relationship between a diverse set of possible risks and impacts. This approach was applied to identify the linkages between the risks derived from climate change, invasive species, and nutrient run-off on the ecological health of the North American Great Lakes (Johnston, 2012).

It is not feasible to undertake detailed assessments of all of potential impacts from climate change discovered during the risk identification process. Instead, the most important risks must be identified for subsequent analysis. For example, the UK Climate Change Risk Assessment used a multi-stage process to produce a long-list of approximately 1 000 conceivable impacts based on expert judgement. This list was subsequently shortened to 100 risks for more detailed analysis (UKCIP, 2013). In some cases, the long list of potential risks can be bundled together, as the risks relating to different sectors all refer to the same underlying potential impact. For instance, when analysing the potential impact of flood risk, property owners view the risk in terms of the potential damages to their properties, the insurance sector by evaluating the likelihood of increased insurance payouts, and the tourism industry in terms of possible changes in the number of visitors to affected tourist attractions. The framing of risks at this identification stage affects how they are characterised and ultimately treated later on in the policy making process.

ii) Characterising risks to be addressed

Having identified a set of risks in the first stage, the next step is to identify for each risk if any changes are necessary to investment in risk reduction or risk transfer mechanisms. The appropriate responses will depend upon multiple factors, such as: the estimated frequency of an event; the severity of its consequences; economic considerations (e.g. the costs and benefits of taking action to reduce the risks); and social dimensions, including distributional impacts. Given the inherent normative dimension of this process, it is important that the criteria adopted are clear, and that there is extensive stakeholder consultation.

Depending on the characteristics of the risk, further information may be needed to determine the preferred combination of: a) risk reduction, b) risk transfer, and c) risk monitoring (Figure 4.3). This categorisation builds on the "risk-layering" approach commonly applied in the context of disaster risk management (e.g. Mechler et al., 2010; Linnerooth-Bayer and Stigler, 2014). These elements are closely linked: incentives to reduce risks can be affected by the risk transfer mechanisms in place. Similarly, each category can involve multiple policy instruments. For example, measures to address excess heat mortality can include land-use management policies, building standards and healthcare provision.

Figure 4.3. **Characterising risks**

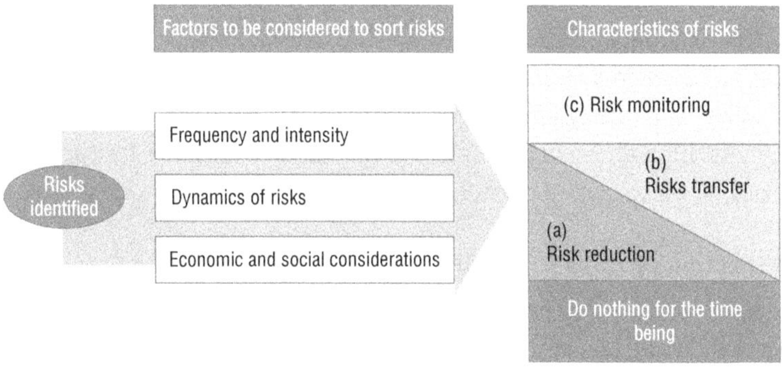

Source: Based on OECD (2013a), *Water Security for Better Lives*, OECD Studies on Water, OECD Publishing; Mechler et. al., (2010) *Assessing the financial vulnerability to climate related natural hazards*, Policy Research Working Paper, 5232, World Bank, Washington, DC.

For many climate related risks, there will already be risk reduction and risk transfer measures in place. For example, levees to reduce the occurrence of flooding and insurance mechanisms to transfer some of the economic losses when a flood occurs. Policies that exist for reasons unrelated to climate change can also change the incidence of climate impacts, for example social safety-net mechanisms. It is necessary to examine whether the current risk management arrangements are satisfactory, both in terms of reducing the risks and transferring any residual impacts. This examination should consider both efficiency and equity dimensions.

Frequency and intensity of adverse consequences

Risk and vulnerability assessments can be used to generate estimates of the projected frequency and consequences of some climate risks (Klinke and Renn, 2012). However, there are inherent limits in modelling climate change and climate impacts, which mean that the likelihood and consequences of some climate risks are uncertain. In the near-term, these are a particular issue for less frequent but high-impact events (Mechler et al., 2010). Moreover, the information on probability and intensity of consequences can be expressed in multiple ways, ranging from quantitative assessment (e.g. commonly used for flood risk management) to more qualitative approaches (e.g. the projected impact on cultural heritage). The breadth of climate risks means that the scoring is likely to be undertaken by different groups of experts and other stakeholders. Common terminology and scoring criteria can help to maintain consistency between these expert judgements.

To compare diverse potential impacts, the OECD's Guidelines for Resilience Systems Analysis recommends rating risks on a scale from "unlikely" to "very likely". Alternatively, they can be rated in terms of their probability of occurring in a given period of time (OECD, 2014b). Such evaluations can, for example, draw on existing contingency plans, national risk assessments, expert analysis and insurance data. At the same time, the guidelines suggest that the scale used for probability assessment should be simple enough for all stakeholders to understand it. An analysis and rating of the impacts of different risks can be informed by quantitative criteria (e.g. the estimated average impact of a particular risk on a specific asset) and qualitative criteria (e.g. expert judgements on severity of risks through interviews with relevant actors) [OECD, 2014b].

Reducing the consequences of recurring low- or moderate-intensity events is an important aspect of risk-based approaches to adaptation planning. These risks tend to develop slowly and can be "invisible" in the short-term, yet their accumulated impacts may cause considerable damage in the long-term (OECD, 2013a). Similarly, some risks comprise a set of smaller risks, each of which may be considered insignificant, but their cumulative effects will not be. Currently, evidence from OECD countries shows that responses to low-frequency and high-intensity risks are prioritised over more-frequent but less severe (EEA, 2014; Groot, Rovisco and Lourenço, 2014).

Dynamic perspective of risks and uncertainties

The uncertain nature of climate change combined with the impact of socio-economic development means that the characteristics of risks will change over time, and the bounds of uncertainty will increase (see Figure 4.4 and Box 4.1). For near-term and medium-term risk management (e.g. up to 30 years), natural climate variability will mask much of the climate signal. Historical experience can therefore be informative about the scale and characteristics of risks. In the longer-term (e.g. 30+ years), however, the application of historic data and experience will be less informative for the range of possible consequences, and greater use of climate models will be needed.

An implication of this is that the likelihood of encountering adverse shocks increases over longer time horizons. As the climate system is increasingly perturbed by human activities, the likelihood of encountering unanticipated adverse shocks increases. Because they occur so infrequently, "tail risks" can have consequences and effects that had not previously been contemplated (OECD, 2014a). As a result, stress testing against different scenarios, designing in flexibility and regular monitoring become particularly relevant over longer time horizons.

The assumptions made about the adaptation responses implemented under a business-as-usual scenario will significantly affect the perceived evolution of risks over time. For example, people will change their behaviour and adapt their buildings to higher temperatures as temperatures increase. In that context, extrapolation of current relationships between

Figure 4.4. **Dynamics of risks over time**

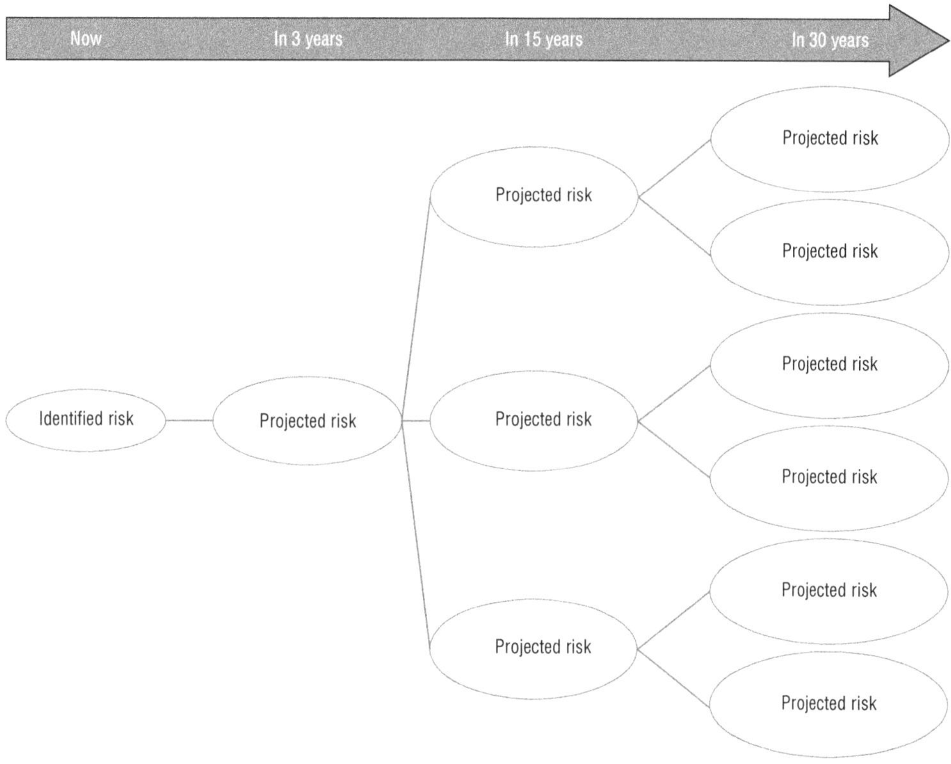

temperatures and excess mortality has the potential to be misleading. This is challenging to model as the relationship will change over time, even in the absence of specific adaptation policies. Factors that can affect autonomous adaptation include: knowledge about climate and non-climate risks, resources and decision-making processes within communities (World Bank, 2014). The evolution of relevant policies is difficult to predict. It is, therefore, important that the assumptions used to inform policy-making processes are transparent and consistent, so that their implications can be understood by decision makers.

Box 4.1. **Drivers of uncertainty**

Uncertainty in climate predictions generally arises from multiple sources. Hawkins and Sutton (2009) outline three key drivers of uncertainty:

- **Internal variability of the climate system:** The natural fluctuations that arise in the absence of any radiative forcing of the planet.
- **Model uncertainty:** In response to the same radiative forcing, different models simulate somewhat different changes in climate. (Also known as response uncertainty)
- **Scenario uncertainty:** Uncertainty in future emissions of greenhouse gases, for example, causes uncertainty in future radiative forcing and hence climate.

The influence of those drivers on the projections of climate change impacts varies with prediction time scales as shown in the figure below. For time horizons of many decades or longer, the dominant drivers of uncertainty are model uncertainty and scenario uncertainty. For time horizons of a decade or two, model uncertainty and internal variability have greater importance on uncertainty. In general, the importance of internal variability increases at smaller spatial scales.

Box 4.1. **Drivers of uncertainty** *(cont.)*

Decadal mean surface air temperature at the global level

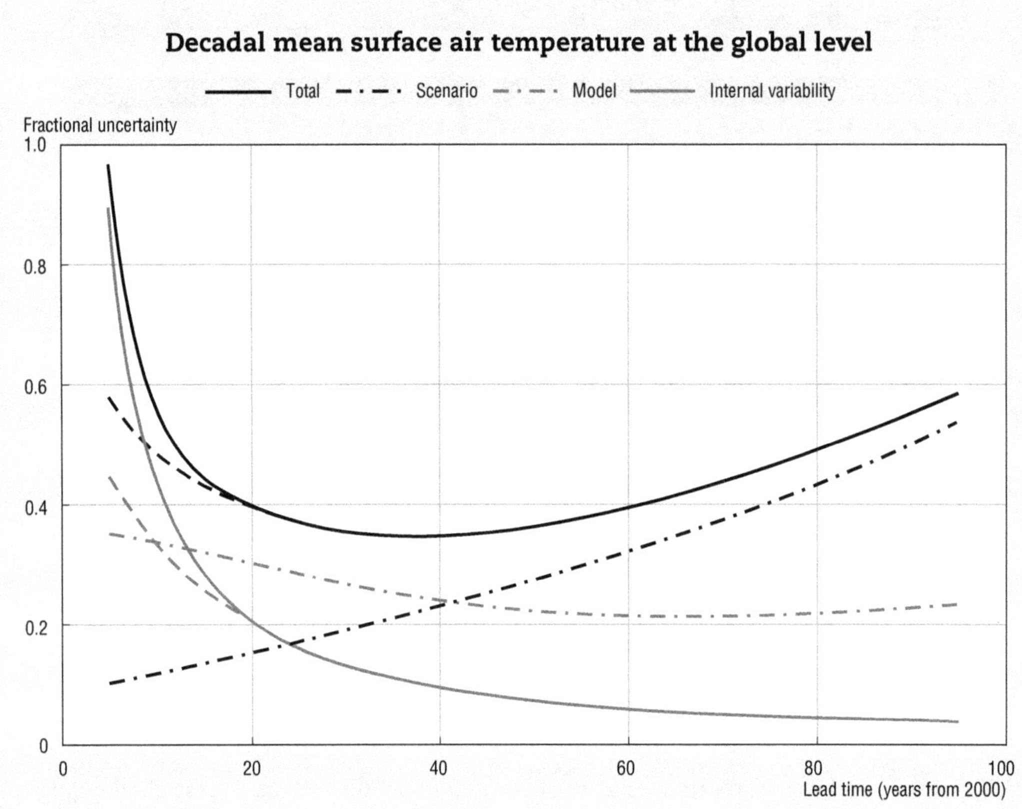

The study also shows that contributions of internal variability and model uncertainty to the uncertainty in projections can be reduced through further development of climate science. Greater uncertainty can increase the cost of addressing climate risks, as adaptation measures have to be designed to perform well under a wider range of potential future climates. Improvements in climate science could help to improve the targeting of resources for adaptation.

Source: Hawkins, E. and R. Sutton (2009), *The potential to narrow uncertainty in regional climate predictions*, Bulletin of the American Meteorological Society, 90(8), 1095-1107 (©American Meteorological Society. Used with permission).

Economic and social dimensions of risk

A risk-based approach to adaptation planning is more than a mechanical process of identifying and sorting risks based on scientific estimates of their projected frequency and intensity. Instead, it should also consider the economic and social implications of the risks, and in particular, how they relate to vulnerable segments of the population. Consideration should also be given to how people that will likely be affected by climate change perceive these risks, recognising that different actors may prioritise the risks and solutions differently. To address this, different stakeholder preferences concerning risk can be developed into agreed principles or criteria for characterising risks. Such principles or criteria can be determined based on stakeholder engagement.

It is also important for relevant stakeholders to agree on a threshold beyond which the consequences of climate-related risks are no longer perceived to be tolerable. This threshold can then provide the basis for identifying when action is needed and what suitable adaptation measures may look like. Social variables, such as age, health, income and mobility can affect opinions on such thresholds. These variables can also influence individuals' ability to prepare for, respond to, and recover from a natural disaster or other potential consequences of climate change (Preston et al. 2014; Cutter et al., 2009; Pastor et al., 2006; Hewitt, 1997).

Incorporating social considerations into the assessment, characterisation and prioritisation of climate-related risks can improve the quality and acceptability of the results. An example is the Social Vulnerability to Climate Change assessment conducted by the state of California in 2012 (Cooley et al., 2012). This assessment included a climate vulnerability index to indicate the social vulnerability of the region's population to climate-related harm. The index used information generated from 19 indicators to calculate an overall climate vulnerability score that included factors specifically related to climate change (Cooley et al., 2012). Another example is Climate Just, a web-based information tool developed in the United Kingdom. The tool aims to identify potential inequalities and disadvantages resulting from differential exposure to climate hazard and social vulnerabilities (Box 4.2).

Box 4.2. **Methodologies to assess and characterise risks**

Climate Just is a web-based tool that highlights which areas in the United Kingdom are vulnerable to the impacts of extreme weather, including flooding and extreme heat. The tool provides visualised data for local authorities and other organisations that work on climate change or with vulnerable communities to support them in responding to climate change. The map below, for example, overlays exposure to river and coastal flooding with social vulnerability to identify areas of "flood disadvantage". The colours indicate the degree of "flood disadvantage", with the darkest shading showing the areas where this is most acute.

Source: Climate Just (2015) Developed by Climate UK on behalf of the Joseph Rowntree Foundation, in association with the University of Manchester and the Environment Agency, *www.climatejust.org.uk/map* (accessed 05 May 2015).

Social dimensions cannot always be readily quantified and will often need to include subjective judgments. This subjectivity can affect the outcome of the risk-characterisation process. Further, these social dimensions can greatly vary over time due to dynamic interactions between values, knowledge, cultures and institutional arrangements (Pelling, 2011; O'Brien, 2012). There are methods to include social dimensions for characterising risks and choosing suitable measures to address them. For instance, multi-criteria analysis (discussed in Chapter 6) is a tool that aims to incorporate not only the cost-benefit profile of risks to be addressed but also other qualitative variables such as co-benefits, ease of implementation, and acceptability to local population (GIZ, 2013).

In addition to social dimensions, information on costs and benefits to characterise risks also involves many subjective choices of variables and uncertainties, and may change over time. Box 4.3 illustrates how OECD member countries are assessing and sorting risks using quantitative and/or qualitative information. It shows that more OECD countries use qualitative methodologies than quantitative methodologies to assess the characteristics of adaptation measures.

Box 4.3. Methodologies to assess and characterise risks

Qualitative tools include expert judgment and stakeholder engagement. Around half of the respondents have used these approaches for setting priorities for the national adaptation strategies or plans. Quantitative tools such as multi-criteria analysis, cost-benefit and cost-effectiveness tools are used by few countries, due partly to limited availability of data.

Techniques used for prioritising the measures to be included in the national adaptation strategies or plans

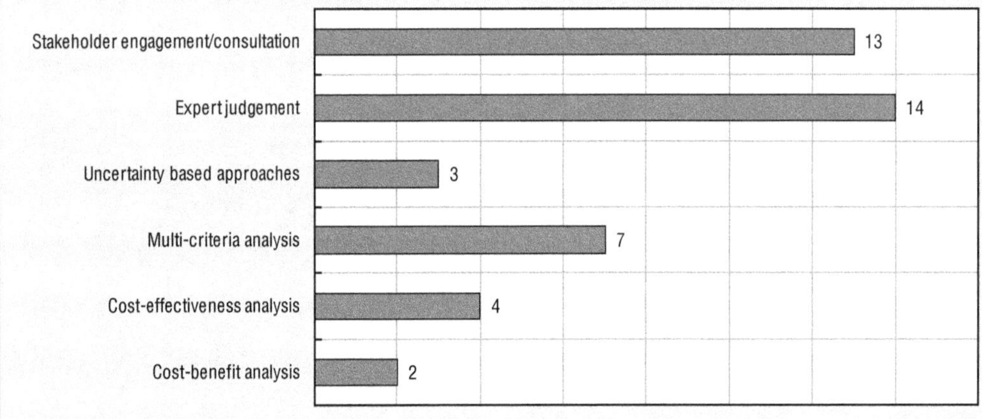

Note: The number shows how many responding countries have applied each technique for developing their national adaptation strategies or plans (Multiple answer).

Links between risk reduction, transfer and absorption

Adaptation policies can support risk management across three strands: reducing risks, transferring risks, or improving resilience to the risks that must be absorbed. These three strands are inter-linked and the distinctions between them are not always clear-cut (see Figure 4.3). As such, multiple policy measures may need to be implemented in a coherent manner across all three categories. Since the characteristics and contexts of risks change over time, it is necessary to periodically assess and (re)characterise the risks and to evaluate what the most suitable measures are.

Risk reduction measures tend to focus on frequently occurring events (Linnerooth-Bayer and Stigler, 2014). Given the greater level of information available on high-frequency risks than on less frequent ones, it is easier for policy makers (as well as the private sector and households) to estimate the costs and benefits of addressing the risks. Possible measures to address such risks include constructing levees against floods, changing crops, implementing early warning systems and developing adaptive water resource management systems (IPCC, 2014).

Even with risk reduction measures in place, there will always be some remaining risks. Both on the basis of equity and efficiency, it can be important to examine how the residual risks are transferred. Risk transfer shifts the risk to others that are in a position to provide compensation if the event (e.g. natural disasters) occurs in exchange for a premium (Poole, 2014). Examples of risk transfer mechanisms include insurance (e.g. crop insurance, weather-index based insurance and national hazard insurance), reinsurance, and catastrophe insurance pools (Linnerooth-Bayer et al., 2010). Chapter 5 discusses some of these risk transfer measures in more detail.

Some consequences of climate change will have to be absorbed, as they cannot be eliminated through risk reduction measures or entirely transferred. This may, for example, be the case with highly uncertain risks or risks with very high environmental and economic consequences. Stress-testing can be a useful tool to identify the threshold beyond which businesses and governments would find it too costly to reduce risks or insure against extreme weather events (Mechler et al., 2010). Finance and insurance industries in the EU typically set the level of 200-year events as their threshold above which risk reduction or transfer measures are no longer cost-effective (EIOPA, 2011). Costs beyond this limit will be absorbed by households, businesses and the public sector. Enhancing resilience can support their ability to absorb those risks.

iii) Choosing and exploring policy options

Once risks have been characterised, suitable risk reduction or risk transfer measures must be identified and implemented. This section examines the types of measures that can be applied, while Chapter 5 focuses on how they can be financed.

A diverse set of approaches implemented by different actors will be needed to effectively address climate risks. For example, in the case of drought, risk reduction measures may entail the introduction of drought-resilient crops, reforms of water allocation policies and the introduction of early warning systems. National or local governments may implement reforms to water abstraction licensing and the introduction of early warning systems, while farmers would choose which crop varieties to use. To complement risk reduction measures, both the public and the private sector can provide risk transfer instruments (e.g. crop insurance) to farmers. Since some residual risks may persist after the mixture of risk reduction and transfer measures have been implemented, governments must be aware of, and prepare for, their contingent liabilities in the event of a catastrophic event. Similarly, private stakeholders may choose to set money aside in case crops fail.

Measures that have been put in place for non-climate reasons can affect the characteristics of climate risks. For example, some distorting agricultural subsidies may inadvertently exacerbate risks by discouraging famers from changing crop types. It may, therefore, not always be necessary to introduce new policy measures to address identified risks. Instead, policy misalignments within existing policy frameworks can be identified, improvements examined and revisions made if necessary.

Adaptation measures can broadly be categorised as "soft" and "hard" (e.g. EEA, 2014). Soft adaptation measures include information provision, capacity building, policy and strategy development and institutional arrangements. Hard adaptation measures imply the use of specific technologies and actions, which involves, for instance, capital goods (e.g. dikes, seawalls and reinforced buildings) [World Bank, 2011]. However, the distinction is not always clear-cut: in practice, soft and hard measures are often used in a combined manner (see Table 4.3).

Table 4.3. **Types of adaptation measures to reduce risks**

Type	Description
"Soft" adaptation	Managerial, legal and policy approaches that aim at altering human behaviour or styles of governance. Examples include early warning systems or knowledge sharing.
"Hard" adaptation	"Grey" measures that aim to reduce vulnerability to climate change and enhance resilience. Examples include building flood defences and beach restoration to prevent coastal erosion. "Green" measures that make use of nature. Examples include introducing new crop and tree varieties, allowing room for rivers to naturally flood onto floodplains, and restoring wetlands.
Combined adaptation	Approaches that use both soft and hard adaptation measures. In fact, the best results are often achieved by combining actions. For example, flood risk in a particular area can be addressed by a combination of "green" and "grey" actions, or "grey" and "soft" actions.

Source: Based on EEA (2013), *Adaptation in Europe – Addressing risks and opportunities from climate change in the context of socio-economic developments*, EEA Report No 3/2013, EEA, Copenhagen.

Some adaptation responses implemented today will have a long life- and lead times (e.g. constructing a seawall or a road), while other responses can be reversed or modified more easily (e.g. changing crop types and knowledge sharing). For investments with long life- and lead times, priority is often given to no-regret adaptation measures that provide benefits independent of climate change. At the same time, the OECD survey demonstrates a strong interest among countries in pursuing synergies between adaptation policies and other policy objectives. In the German Strategy for Adaptation to Climate Change, a major component is on cross-cutting linkages between adaptation policies and other national strategic processes (BMU, 2008).

However, since all risks cannot be addressed by low-cost or no-regret options, there will be areas where significant investments are being made even when the outcomes are uncertain. The Dutch government recently announced a new project to construct sea-walls and dikes that is expected to span 30 years and cost EUR 20 billion (Delta Programme Commissioner, 2014). New decision-making tools (e.g. real option analysis, robust decision making and portfolio analysis discussed in Chapter 6) are being applied to account for uncertainty about the future. These approaches can be used to build in flexibility and reduce costs when investing in long-lived adaptation measures.

Financial measures used to transfer risks include, among others, insurance policies, and capital market instruments such as catastrophe bonds and weather derivatives. These can be provided in different manners at various levels, such as national (e.g. insurance pools and contingent credit), business (e.g. re-insurance and diversification) and household (e.g. savings and credit) levels (see Chapter 5 for further discussion).

Insurance mechanisms are one of the most prevalent risk transfer mechanisms in OECD countries, with private insurance covering around 40% of weather-related damages (Warner et al., 2012). Partnerships between private and public insurers can help to increase the coverage of insurance where climate risks are difficult to insure at affordable rates. Measures other than insurance can also have an important role in managing climate risks. Such measures can include disaster funds, social safety nets and support for recovery of local industries after disasters that can help strengthen the long-term efficiency of adaptation measures and equity across different social groups and sectors.

Policy makers need to consider possible trade-offs between risk reduction and risk transfer. Insurance can lead to moral hazard, as policyholders have less incentive to reduce risks than they would if they bore the risks themselves. This can be mitigated by the use of risk-based premiums and deductibles. Similar issues affect disaster relief payments and ex-post compensation. The design of risk transfer mechanisms can ameliorate this by encouraging or requiring policyholders to adopt appropriate risk-management behaviour or by simply demanding it (IPCC, 2012; Surminski, 2010). For example, index-based insurance that is determined by climate variables and not actual losses can be effective in creating incentives for policyholders to reduce their risks (OECD, 2008).

iv) Feedback process

Regular monitoring and evaluation are essential components of climate risk management. Long time-horizons and uncertainty about the future mean that risk management needs to be flexible and iterative (Wise et al., 2014; Haasnoot et al., 2013). Monitoring and evaluation provide a set of tools to achieve this by assessing whether assumptions about the characteristics of risks need to be updated in light of new information. OECD (2015) shows how four tools can be combined to achieve this at the national level: climate change risk and vulnerability assessments, indicators to monitor progress on adaptation priorities, project and programme evaluations to identify effective adaptation approaches, and national audits and climate expenditure reviews.

Lessons learned from monitoring and evaluation should be used to inform adjustments to the adaptation planning process (OECD, 2015; Dinshaw et al., 2015). Any such adjustments should consider the dynamic interactions between social dimensions such as values; knowledge, cultures and institutions that enable or constrain decision-making processes; as well as the evolving nature of risks (Wise et al., 2014). Figure 4.5 outlines a conceptual framework for the feedback process and provides questions that could be used to inform possible adjustments.

The design and implementation of feedback mechanisms should be considered from the outset of the risk-based process. This ensures that the mechanisms provide the right information at the right time to inform decision making. An example of this is provided by the Thames Estuary 2100 project (flood defences for London), which identified monitoring requirements and decisions points throughout the life of the project. At the strategic level, the legislative framework for climate policy in the United Kingdom (Climate Change Act [2008]) synchronises the development and revision of adaptation plans with monitoring and evaluation. Institutional arrangements such as these, accompanied with political will, can help to ensure that monitoring and evaluations are integrated into decision making.

Figure 4.5. **Conceptual framework of a feedback process for a risk-based approach**

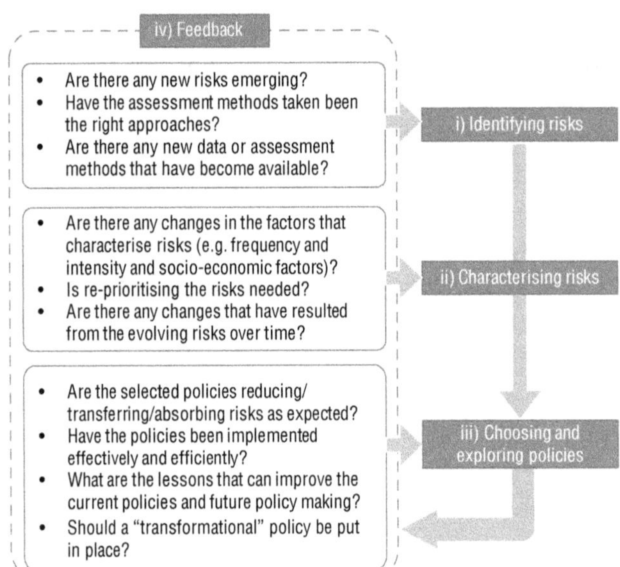

References

Ayers, J. and T. Forsyth (2009), *Community-based adaptation to climate change: Strengthening resilience through development*, *Environment*, Vol. 51(4).

BMU (The German Federal Ministry for the Environment, Nature Conservation and Nuclear Safety) (2008), *German Strategy for Adaptation to Climate Change*, BMU, Berlin.

Cabinet Office (2013), *Review of cross-government horizon scanning*, *www.gov.uk/government/publications/review-of-cross-government-horizon-scanning* (Accessed 19 February 2015).

Chmutina, K. et al. (2014), *Towards Integrated Security and Resilience Framework: A Tool for Decision-makers*, Procedia Economics and Finance, Vol. 18, *http://dx.doi.org/10.1016/S2212-5671(14)00909-5*.

Consorcio de Compensación de Seguros (2012), *The 'Consorcio de Compensación de Seguros' and the Spanish Natural Catastrophe Coverage System*, *http://www.wfcatprogrammes.com/c/document_library/get_file?folderId=13442&name=DLFE-553.pdf* (accessed 19 February 2015).

Cooley, H. et al. (2012), *Social Vulnerability to Climate Change in California*, California Energy Commission, Pacific Institute, Oakland.

Cutter, S.L. et al. (2009), *Social Vulnerability to Climate, Variability Hazards: A Review of the Literature*, Final Report to Oxfam America, Hazards and Vulnerability Research Institute and University of South Carolina.

Dinshaw, A. et al. (2015), *Monitoring and Evaluation of Climate Change Adaptation: Methodological Approaches*, OECD Environment Working Papers, No. 74, *http://dx.doi.org/10.1787/5jxrclr0ntjd-en*.

Delta Programme Commissioner (2014), *2014 – the year of the new Delta Plan*, Delta Programme Commissioner, *http://english.deltacommissaris.nl/news/news/2014/12/22/2014-%E2%80%93-the-year-of-the-new-delta-plan* (accessed 12 February 2015).

Dessai, S. and M. Hulme (2004), *Does Climate Adaptation Policy need Probabilities?*, Climate Policy, Vol. 4(2).

EC (2014), *The economic impact of climate change and adaptation in the Outermost Regions: Final report*, European Commission, Directorate-General for Regional and Urban policy, Brussels.

EC (2009), *Impact Assessment Guidelines*, European Commission, Brussels.

EEA (2014), *National adaptation policy processes in European countries: 2014*, European Environment Agency, Copenhagen.

EEA (2013), *Adaptation in Europe – Addressing risks and opportunities from climate change in the context of socio-economic developments*, EEA Report, No. 3/2013, European Environment Agency, Copenhagen.

EIOPA (2011), *Report on the Fifth Quantitative Impact Study (QIS5) for Solvency II*, European Insurance and Occupational Pensions Authority, Frankfurt.

Füssel, H.-M. (2010), *Review and quantitative analysis of indices of climate change exposure, adaptive capacity, sensitivity, and impacts*, Background note to the World Development Report, Potsdam Institute for Climate Impact Research, Potsdam.

GIZ (2013), *Economic approaches for assessing climate change adaptation options under uncertainty: Excel tools for Cost-Benefit and Multi-Criteria Analysis*, Deutsche Gesellschaft für Internationale Zusammenarbeit (GIZ), Eschborn.

Gorddard, R. et al. (2012), *Striking the balance: Coastal development and ecosystem values*, Department of Climate Change and Energy Efficiency, Canberra.

Government of Japan (2011), *Higashi nihon daishinsai ni taisho suru tame ni hitsuyō na zaigen no kakuho o hakaru tame no tokubetsu sochi ni kansuru hōritsu [Act on Special Measures to Secure a Source of Funds Necessary to React to the Aftermath of the Great East Japan Earthquake]*, Act No. 42 of 2011.

Groot, A., A. Rovisco, and T.C. Lourenço (2014). *Showcasing Practitioners' Experiences*, in Lourenço, T.C. et al. (eds.), Adapting to an Uncertain Climate, Springer International Publishing, Switzerland.

Haasnoot, M. et al. (2013), *Dynamic adaptive policy pathways: A method for crafting robust decisions for a deeply uncertain world*, Global Environmental Change, Vol. 23(2), *http://dx.doi.org/10.1016/j.gloenvcha.2012.12.006*.

Hammill, A. and T. Tanner (2011), *Harmonising Climate Risk Management: Adaptation Screening and Assessment Tools for Development Co-operation*, OECD Environment Working Papers, No. 36, OECD Publishing, Paris, *http://dx.doi.org/10.1787/5kg706918zvl-en*.

Hewitt, K. (1997), *Regions of Risk: A Geographical Introduction to Disasters*, Addison Wesley Longman, London.

IPCC (Intergovernmental Panel on Climate Change) (2014), *Climate Change 2014: Impacts, Adaptation, and Vulnerability*, Contribution of Working Group II to the Fifth Assessment Report of the Intergovernmental Panel on Climate Change, Cambridge University Press, Cambridge and New York.

IPCC (2012), *Managing the Risks of Extreme Events and Disasters to Advance Climate Change Adaptation*, A Special Report of Working Groups I and II of the Intergovernmental Panel on Climate Change, Cambridge University Press, Cambridge and New York.

JICA (Japan International Cooperation Agency) (2013), *Response to the Great East Japan Earthquake: Overview and Lessons Learned from the Great East Japan Earthquake*, *http://www.jica.go.jp/english/publications/reports/annual/2013/c8h0vm00008m8edo-att/62.pdf* (accessed 12 February 2015).

Johnston, E. (2012), *Seeing a Bigger Picture in the Great Lakes: Systems Mapping for Powerful Insights*, Climate Interactive, *www.climateinteractive.org/analysis/seeing-a-bigger-picture-in-the-great-lakes-systems-mapping-for-powerful-insights/* (accessed 15 January 2015).

Joseph Rowntree Foundation (2015), *JRF DATA: River and costal flood exposure*, *http://data.jrf.org.uk/data/river-and-coastal-flood-exposure/* (accessed 24 January 2015).

Kates, R.W., W.R Travis and T.J. Wilbanks (2012), *Transformational adaptation when incremental adaptations to climate change are insufficient*, Proceedings of the National Academy of Sciences of the United States of America, Vol. 109(19), *http://dx.doi.org/10.1073/pnas.1115521109*.

Keskitalo, E.C.H., G. Vulturius and P. Scholten (2013), *Adaptation to Climate Change in the Insurance Sector: Examples from the UK, Germany and the Netherlands*, Natural Hazards, Vol. 71(1), *http://dx.doi.org/10.1007/s11069-013-0912-7*.

Klinke, A. and O. Renn (2012), *Adaptive and integrative governance on risk and uncertainty*, Journal of Risk Research, Vol. 15(3), *http://dx.doi.org/10.1080/13669877.2011.636838*.

Linnerooth-Bayer, J. and S. Hochrainer-Stigler (2014), *Financial instruments for disaster risk management and climate change adaptation*, Climatic Change, *http://dx.doi.org/10.1007/s10584-013-1035-6*.

Linnerooth-Bayer, J., R. Mechler and S. Hochrainer-Stigler (2010), *Insurance against Losses from Natural Disasters in Developing Countries: Evidence, gaps and the way forward*, IDRiM Journal, Vol. 1(1), *http://dx.doi.org/10.5595/idrim.2011.0013*.

Mechler, R. et al. (2010), *Assessing the financial vulnerability to climate related natural hazards*, Policy Research Working Paper, No. 5232, World Bank, Washington, DC.

O'Brien. K. (2012), *Global environmental change II: From adaptation to deliberate transformation*, Progress in Human Geography, Vol. 36 (5), *http://dx.doi.org/10.1177/0309132511425767*.

OECD (Organisation for Economic co-operation and Development) (forthcoming), *National Climate Change Adaptation: Emerging Practices in Monitoring and Evaluation*, OECD Publishing, Paris, *http://dx.doi.org/10.1787/9789264229679-en*.

OECD (2014a), *Boosting Resilience through Innovative Risk Governance*, OECD Publishing, Paris, *http://dx.doi.org/10.1787/9789264209114-en*.

OECD (2014b), *Guidelines for Resilience Systems Analysis*, The Development Assistance Committee, Enabling Effective Development, *www.oecd.org/dac/Resilience%20Systems%20Analysis%20FINAL.pdf* (accessed 12 February 2015).

OECD (2013a), *Water Security for Better Lives*, OECD Studies on Water, OECD Publishing, Paris, *http://dx.doi.org/10.1787/9789264202405-en*.

OECD (2013b), *Water and Climate Change Adaptation: Policies to Navigate Uncharted Waters*, OECD Studies on Water, OECD Publishing, Paris, *http://dx.doi.org/10.1787/9789264200449-en*.

OECD (2012), *OECD Environmental Outlook to 2050: The Consequences of Inaction*, OECD Publishing, Paris, *http://dx.doi.org/10.1787/9789264122246-en*.

OECD (2008), *Economic Aspects of Adaptation to Climate Change: Costs, Benefits, and Policy Instruments*, OECD Publishing, Paris, *http://dx.doi.org/10.1787/9789264046214-en*.

OECD and FAO (2012), *Building resilience for adaptation to climate change in the agriculture sector*, Proceedings of a Joint FAO/OECD Workshop, 23-24 April 2012, *www.fao.org/docrep/017/i3084e/i3084e.pdf* (accessed 12 February 2015).

OECD/G20 (2012), *Disaster Risk Assessment and Risk Financing: A G20/OECD Methodological Framework*, OECD, Paris.

Park, S.E. et al. (2012), *Informing adaptation responses to climate change through theories of transformation*, Global Environmental Change Vol. 22(1), *http://dx.doi.org/10.1016/j.gloenvcha.2011.10.003*.

Pastor. M. et al. (2006), *In the Wake of the Storm: Environment, Disaster and Race After Katrina*, Russell Sage Foundation, New York.

Pelling, M. (2011), *Adaptation to Climate Change: From Resilience to Transformation*, Routledge, London.

Preston, I. et al. (2014), *Climate Change and Social Justice: An Evidence Review*, Joseph Rowntree Foundation, York, UK.

Poole, L (2014), *A Calculated Risk: How Donors Should Engage with Risk Financing and Transfer Mechanisms*, OECD Development Co-operation Working Paper, No. 17, OECD Publishing, Paris, *http://dx.doi.org/10.1787/5jz122cn65s6-en*.

Seifert, I. et al. (2013), *Influence of flood risk characteristics on flood insurance demand: A comparison between Germany and the Netherlands*, Natural Hazards Earth System Sciences, Vol. 13, *http://dx.doi.org/10.5194/nhess-13-1691-2013*.

Surminski, S. (2010), *Adapting to Extreme Weather Impacts of Climate Change – How can the Insurance Industry Help?* Climate Wise, London.

Sutherland, W.J. and H.J. Woodroof (2009), *The need for environmental horizon scanning*, Trends in Ecology & Evolution, Vol. 24(10), *http://dx.doi.org/10.1016/j.tree.2009.04.008*.

UKCIP (UK Climate Impacts Programme) (2013), *Dialogue on National Climate Change Assessments: Summary Report*, summary of a dialogue meeting that took place in London, UK, 6-7 February 2013, *www.ukcip.org.uk/wordpress/wp-content/national-assessments/2013-May-summary-report-international-dialogue.pdf* (accessed 12 february 2015).

UNCCD (2009), *LAND: A tool for climate change adaptation*, UNCCD Policy Brief, No. 1, United Nations Convention to Combat Desertification, Bonn.

Warner, K. et al. (2013), *Innovative Insurance Solutions for Climate Change: How to integrate climate risk insurance into a comprehensive climate risk management approach*, MCII Policy Brief No. 12, United Nations University Institute for Environment and Human Security (UNU-EHS), Bonn.

Warner, K. et al. (2012), *Insurance solutions in the context of climate change-related loss and damage: Needs, gaps, and roles of the Convention in addressing loss and damage*, MCII Policy Brief No. 6, United Nations University Institute for Environment and Human Security (UNU-EHS), Bonn.

Wise, R.M. et al. (2014), *Reconceptualising adaptation to climate change as part of pathways of change and response*, Global Environmental Change, Vol. 28, *http://dx.doi.org/10.1016/j.gloenvcha.2013.12.002*.

World Bank (2014), *Tajikistan – Autonomous adaptation to climate change: Economic opportunities and institutional constraints for farming households*, Report No. 85597-TJ, World Bank, Washington, DC.

World Bank (2011), *Key Words and Definitions*, World Bank, *http://web.worldbank.org/WBSITE/EXTERNAL/TOPICS/ENVIRONMENT/EXTTOOLKIT3/0,,contentMDK:22284629?pagePK:64168445?piPK: 64168309?theSitePK:3646251,00.html* (accessed 23 October. 2014).

WWF (World Wildlife Fund) and RSA (2014), *Environmental Systemic Risk & Insurance White Paper*, WWF-UK and RSA Group.

Climate Change Risks and Adaptation
Linking Policy and Economics

Chapter 5

Financing adaptation in OECD countries

The chapter examines how OECD countries can finance adaptation to manage climate-related risks. It first analyses the financing of risk reduction investments and risk transfer mechanisms. It also discusses possible options for governments to absorb the liabilities arising from residual risks. Furthermore, the chapter explores the role of governments in supporting the further uptake of financial instruments at the national, sub-national and sectoral levels.

Key messages

- Climate risks will have direct financial implications for governments, but also give rise to indirect effects. The development of financing strategies, linking risk reduction and risk transfer, can help to ensure an efficient response and clarify expectations.
- Data on the costs of climate risks are limited, which can inhibit adaptation planning. A better understanding of the fiscal impact of climate risks – damage to assets, compensation payments, reduction in tax revenues – would support risk management. Identifying and reporting potential contingent liabilities from climate change could inform measures to address those liabilities
- Governments can, and need to, play a significant role at various levels (i.e. national, sub-national and sectoral) in enabling finance for adaptation, both directly through public sector expenditures, and indirectly by facilitating private sector action.

Framework for understanding the budgetary impacts of climate change adaptation

The impacts of climate change are likely to lead to net economic costs. The adaptation financing strategies adopted by governments will determine both the eventual scale of these costs and how they are distributed. There are three main channels through which government finances will be affected by climate change:

- **Investments to support risk reduction:** The allocation of resources to reduce the probability and/or consequences of a climate risk. For example, considering exposure to climate change risks when deciding where to locate infrastructure, or preparing contingency plans for heat waves. Investments in risk reduction can entail implementing things differently rather than necessarily at higher cost.
- **Risk transfer mechanisms:** This term is defined broadly to include mechanisms that provide monetary compensation when a climate-related event occurs. These include insurance, disaster relief payments and financial market instruments, such as catastrophe bonds.
- **Absorption of residual impacts:** Governments will be directly affected by impacts on operations and maintenance of public sector buildings due to trend changes, damage to public infrastructure or through payments mandated under risk transfer mechanisms. Government finances may also be indirectly affected if, for example, changes in trade flows affect tax revenues or social welfare payments.

The three financing channels listed above are interrelated. Investment in risk reduction should yield reductions in the scale of risks to be absorbed. Improving awareness of the potential risks to be reduced can strengthen the case for investments in risk management (Dellink et al., 2014; IEG, 2013; Israel, 2013). In practice, however, the relationship between risk reduction and risk absorption is difficult to monitor. Governments only routinely track some of their direct expenditures on risk reduction (e.g., flood defences) and risk absorption. Governments do not routinely track indirect costs arising from climate risks, yet these costs could be significant. The scale of finance required for risk absorption will depend upon the scale and effectiveness of both public and private investments in adaptation. Institutions, incentives and regulations will affect the scale of private sector investment (Agrawala et al., 2012). In addition, the baseline for comparison is not static. Increasing investment in risk reduction may be insufficient to offset increases in the underlying risks driven by climate change and socio-economic trends. Lastly, the consequences of underinvestment in risk reduction may be invisible for long periods of time until an extreme event occurs.

A key element of the relationship between risk reduction and risk absorption is the extent to which risks are transferred between stakeholders. These include formal risk transfer mechanisms such as agricultural insurance, but also ad-hoc measures to compensate those affected when a risk materialises: for example, temporary tax relief for farmers affected by drought. The consequence of this is that the people directly affected by a risk are not necessarily those who ultimately end up having to absorb the consequences. The existence and design of ex-ante risk transfer mechanisms, and people's expectations about ad-hoc relief, can also affect the incentives of both public and private sector actors to adapt.

A vivid illustration of the scale of risks to be absorbed was provided by the 2013 European floods when the upper Danube basin, the Inn, the Elbe and other rivers overflowed their banks. These floods contributed to an estimated EUR 17 billion in property damages, of which only around EUR 4 billion were insured (Zurich Insurance Group, 2014). In Germany, where flood losses were estimated to be around EUR 12 billion, insurance penetration was at around 35%. The German government made available EUR 8 billion in flood relief to compensate some of the uninsured losses suffered by homeowners and businesses. The government also incurred the costs of repairing damages to levies and other vital infrastructure such as roads (Zurich Insurance Group, 2014). Extreme floods in Europe, like those experienced in 2013, are projected to increase from once every 16 years to once every 10 years by 2050 (Jongman et al., 2014).

Investment to support climate risk reduction

Difficulties in channelling sufficient investment to climate risk reduction are increasingly recognised as a constraint on the implementation of adaptation at the national level (Mullan et al., 2013). The scale of resources required at the national level to reduce the risks from climate change is uncertain, but estimates provided in Chapter 3 suggest that the overall scale of resources will be manageable relative to the size of OECD countries' economies. However, these investment needs will not be evenly distributed by sector or geographical region. Institutional constraints and market failures can prevent financial resources flowing to the areas where they are most needed (Cimato and Mullan, 2010; Agrawala et al., 2012).

The diverse nature of the risks arising from climate change is reflected in the diversity of potential investments that could be used to address those risks. Potential investments include the provision of protective infrastructure, capacity building, climate research, policy research, and subsidies for private adaptation measures. Investment needs also include the costs of reversing the impact of previous policy choices, for example, by supporting households' relocation from high-risk areas (see Box 5.1) or buying back

Box 5.1. Hurricane Sandy: Example of a policy response to manage weather risks

When Hurricane Sandy hit the east coast of the United States in October 2012, it damaged or destroyed around 650 000 homes, killed at least 159 people, temporarily displaced 23 000 people, and left around 8.5 million customers without electricity, some for up to three weeks. In the aftermath of this disaster, the State of New York introduced the Home Buyout Program as part of a larger state programme designed to strengthen and upgrade critical state systems and infrastructure. The objective of the buyback programme is to permanently transform the state's coastal zones into preservation land by spending as much as USD 400 million to buyback homes in high-risk areas.

The Home Buyout Program is a state-sponsored initiative available on a voluntary basis to owners of homes located within the 500-year floodplain that were damaged beyond 50% of their value. The programme offers qualifying homeowners 100% of the pre-storm fair market value of their homes. To maximize participation, additional incentives are available for homeowners that relocate within the same county (an additional 5% of pre-storm fair market value), homes in high-risk areas (additional 10%), and for communities that collectively agree to relocate (additional 10%). There is, however, an upper limit linked to the median home value in a given neighbourhood. As required by the US Department of Housing and Urban Development, the land purchased through the buyback programme must remain publically owned open space (e.g. wetlands, dunes and parkland) to absorb floodwaters and act as coastal buffers. New York, however, is seeking a waiver to this rule to redevelop properties outside the 100-year floodplain that are properly mitigated against future flooding.

Initial response to the programme has been mixed. It has been well received by homeowners that have suffered repeated damages from storms and flooding and are, in turn, facing increasing insurance premiums and reduced property values. However, homeowners that have lived in the area for decades or generations or have already made significant investments in their waterfront homes and lifestyles are less enthusiastic about the programme. If the New York state government succeeds in buying back thousands of homes, changing large stretches of land from urban residential areas to open green spaces, the programme is expected to significantly change the vulnerability of the area to future floods and storm surges.

Source: Binder, S.B. (2013), *Resilience and Post-disaster Relocation: A Study of New York's Home Buyout Plan in the Wake of Hurricane Sandy*, Report prepared for Quick Response Grant Program, Natural Hazards Center, University of Colorado.

abstraction permits. These climate risk reduction investments may sometimes be standalone projects, but will predominantly come in the form of adjustments to projects that would have been undertaken in any case: for example, changing the location or design standards of new construction to reduce exposure to climate risk.

The majority of OECD countries take an integrated approach to adaptation planning and implementation, with only a few countries earmarking funds for adaptation-specific initiatives (Mullan et al., 2013). Under an integrated approach, the scale of finance for adaptation cannot be readily distinguished from other funding streams. The exceptions to this are the relatively small expenditures directly linked to the funding of central co-ordination mechanisms for adaptation and the provision of tools and evidence, such as climate projections. In OECD countries, the majority of the resources required to increase climate adaptation will flow at sectoral level. This makes it difficult, if not impossible, to separate out the risk reduction component of specific interventions.

A few countries have reserved or identified national resources specifically for the implementation of key risk reduction activities, or strengthening adaptive capacity. For example, Canada allocated a total of CAD 149 million (EUR 104 million) from the national budget for adaptation over 2011-16, while France has estimated that EUR 171 million will be required to implement their adaptation plan (Mullan et al., 2013). These funds have focussed primarily on enhancing the enabling environment for adaptation, rather than the implementation of specific measures to reduce risks or exploit opportunities.

The European Commission has an overall target for spending on efforts to reduce the risks from climate change, as well as several earmarked funds. The Commission has committed to spending 20% of its EUR 960 billion budget for 2014 and 2020 on climate change-related action. Within this envelope, funds have been made available specifically for adaptation measures through, for example, the Financial Instrument for the Environment (LIFE) where adaptation projects can be co-financed with EUR 190 million during the period 2014-20 (EC, 2014). Under the LIFE instrument the Natural Capital Finance Facility aims at leveraging private investment for ecosystem-based adaptation projects, as well as biodiversity protection that could generate adaptation benefits as well (e.g. green infrastructure).

Aside from EU funding mentioned above, OECD countries rely predominantly upon domestic resources to finance risk reduction. Mexico, Chile, Korea and Israel are the only OECD countries that are recognised by the UN as Non-Annex I Parties to the Convention on Climate Change. This makes them eligible for financial support from relevant bodies and institutions established under the Convention. Chile and Mexico, for example, are eligible to access the Adaptation Fund. This enables accredited national and regional implementing entities to directly access finance for adaptation initiatives. Although Turkey is listed as an Annex I country, it has been recognised that "the provision of financial, technological and capacity-building support to Turkey is important to assist in the implementation of the Convention" (UNFCCC, 2013).

Although adaptation and disaster risk management are inherently linked, they tend to have different institutional arrangements and funding streams. Under the Sendai Framework for Disaster Risk Reduction (the successor to the Hyogo Framework for Action), countries commit to providing sufficient financing for disaster risk reduction. As much of the funding for disaster risk reduction is also mainstreamed, the evidence on the overall scale of funding is incomplete. Some OECD countries have standalone funds that contribute to disaster risk management. For example, Japan has allocated EUR 3.85 billion (JPY 530.4 billion) to its fund for disaster risk reduction (UNISDR, 2014). Austria has earmarked a proportion of its Federal Disaster Fund to strengthen preparedness for extreme events (OECD, 2013c).

In addition to government resources being allocated for adaptation, various policy instruments can also encourage, or in some cases require, investments in climate resilience by households and businesses (see a few examples summarised in Table 5.1). Market-based instruments, such as rebates for investments in water-efficient equipment, encourage individual stakeholders to adjust their operations or investment decisions when profitable to do so (Elrick-Barr et al., 2014). The application of economic instruments may be paired with regulatory measures that establish a minimum acceptable level of performance. For example, the implementation of land-use planning standards or building regulations can help to direct investment towards lower-risk areas. In addition, policies such as the reform of agricultural policies and regulation of the water industry influence the private sector's willingness and ability to invest.

Table 5.1. **Tools for government action to support private adaptation**

Category of government action	Purpose	Examples
Assessment, measurement, reporting and verification	Demonstrate a commitment to evidence-based and transparent adaptation investment. Demonstrate the logic of public investment and attract additional funding.	Publication of vulnerability, risk, or adaptive capacity assessments. Tracking and public reporting of adaptation performance indicators.
Regulation, plans, and policies	Raise and maintain confidence that the operating environment of a given sector (e.g. land use) will be consistent. Demonstrate a commitment to climate change adaptation and an ability to deliver service effectively. Develop a culture of proactive leadership and innovation.	Climate-smart policies to influence private sector activity, ranging from stricter land use administration to guide development away from vulnerable floodplain lands, to positive incentives to promote green infrastructure among building and infrastructure developers.
Fiscal incentives	Cover the incremental costs of adaptation (e.g. building a stronger foundation for a facility already under construction). Encourage investments primarily dedicated to adaptation (e.g. increasing the elevation of existing buildings in zones exposed to frequent flooding).	Tax benefits, subsidies, property taxes differentiated by risk, differentiated insurance premiums, subsidised loans. Provision of cash payments for home renovations that reduce vulnerability to motivate some homeowners, especially if viewed as a time-limited opportunity.
Inducement prizes and public recognition of corporate responsibility	Promote excellence and leadership by example among private sector actors.	Green building ratings. Corporate sustainability awards.

Source: World Bank (2011), *Guide to Climate Change Adaptation in Cities*, World Bank, Washington, DC.

Climate risk transfer

Risk transfer mechanisms are an important complement to investment in risk reduction, as it will rarely be cost-effective or technically feasible to reduce risks to zero. There is also the closely related concept of risk sharing, when risks are pooled within a group. Risk transfer mechanisms affect both the distribution of impacts within the private sector, and also the share that is ultimately absorbed by the public sector. Climate change will put pressure on the provision of private risk transfer due to the rising scale of losses and the uncertainty in the modelling of how those losses will evolve. This could lead to increased demands for governments to bear some of the risks, either through direct provision or by taking on some of the risks that would otherwise be held by private sector insurers.

Insurance arrangements have a key role to play in the transfer of risks resulting from climate change. Insurance can also facilitate recovery and development after extreme events and reduce longer-term indirect losses (see Table 5.2 for some examples) (Bräuninger et al., 2011). A general challenge with insurance mechanisms is to ensure that the transfer of risks does not reduce the incentives for policy holders to undertake risk reduction activities. Deductibles, policy requirements and coverage restrictions can all be used to encourage risk reduction by the policy holders (Surminski, 2010). Further, when insurance costs reflect the real level of risk, they can also incentivise risk reduction behaviour (e.g. by discouraging risky behaviour and rewarding risk reduction).

There are, however, barriers that can prevent insurance premiums from reflecting the severity of the underlying risk. The first is that the level of risk will depend upon the risk reduction measures in place, as well as the value of assets exposed to the risk. In practice, it is too time and resource intensive for insurance providers to undertake the detailed risk assessment required to ensure that premiums fully account for these factors (Bräuninger et al., 2011). Added to these practical constraints, legislative or regulatory requirements may prevent insurers from setting premiums that reflect the underlying risks.

Table 5.2. **Climate change projections of insured losses and/or insurance prices**

Hazard	Insurance line	Region	Projected changes in future time slices relative to current climate (spatial distribution and vulnerability of insured values assumed to be unchanged over time)
Winter storm	Homeowners insurance	Europe	• Although storm track behaviour is poorly characterised in climate models, projected increases in mean annual loss ratio in e.g. France, Belgium, Netherlands, the United Kingdom, Ireland, Germany, and Poland lie in a range of 1-to-2 digit percentages before and around 2050, with larger increases at the end of the century. In Southern Europe (e.g. Portugal and Spain) losses are expected to decrease. • Today's 20-year, 10-year, and 5-year return periods will for all of Europe be roughly halved by the end of the century.
River flood, marine flood, flash flood from rainfall, melting snow	Property and business interruption insurances	Europe, North America	• **Germany:** Projected increases in mean annual insured flood loss are 84% (2011-40), 91% (2041-70) and 114% (2071-2100). • **The United Kingdom:** Projected increases in mean annual insured flood loss are 8% for a 2 °C global mean temperature rise and 14% for a 4 °C rise, with the 1-in-100-year loss higher by 18% and 30%, respectively. • **Canada:** Losses from heavy precipitation in property and business interruption insurances in three city areas are projected to rise by 13% (2016-35), 20% (2046-65), and 30% (2081-2100). • **Norway:** In three counties across southern Norway, precipitation and snow melt insurance losses are expected to be higher by approximately 10-21% and 17-32% at the end of the century. • **The Netherlands:** Expected annual property loss caused by increasing river discharge and sea level rise with an assumed flood insurance system is projected to lie 125% higher in 2040 relative to 2015 (with a 24 cm sea level rise) and by 1784% higher in 2100 (85 cm sea level rise).
Tropical cyclone	Foremost property insurance lines	North America, Asia	• **The United States:** The price levels of Florida's hurricane wind insurance are projected to change by –20% to +5% (2020s) and –28% to +10% (2040s) (under the assumptions of strained reinsurance capacity). • **China:** Projected increases of insured typhoon losses are 20% (for a 2 °C scenario) and 32% (for a 4 °C scenario), with the 1-in-100-year loss higher by 7% and 9%, respectively.
Hailstorm	Homeowners' insurance, agricultural insurances	Europe	• **The Netherlands:** Losses from outdoor farming insurance and greenhouse horticulture insurance are projected to increase by 25-29% and 116-134%, respectively, in a 1 °C scenario. For a 2 °C scenario, projected increases are 49% to 58% and 219% to 269%, respectively. • **Germany:** Projected increases in mean annual loss ratios from homeowners' insurance due to hail are 15% (2011-40) and 47% (2041-70).
Storms, pests, diseases	Paddy rice insurance	Asia	• **Japan:** Paddy rice insurance payouts are projected to decrease by 13% by the 2070s, on the basis of changes in standard yield and yield loss.

Source: Adapted from IPCC (2014a), *Climate Change 2014: Impacts, Adaptation and Vulnerability*, Contribution of Working Group II to the Fifth Assessment Report of the Intergovernmental Panel on Climate Change, Field, C.B. et al. (eds.), Cambridge University Press, Cambridge, UK and New York, NY, US.

Reconciling the equity and efficiency objectives of insurance arrangements is a key challenge for policy makers. One aspect of this has been the subsidisation of those living in higher-risk areas, either explicitly or implicitly through premiums that do not fully reflect the underlying risks. The costs of this are either borne by other policy holders or by the state. If the cross-subsidy element is reduced then those households living in high-risk areas could see marked increases in their premiums. Box 5.2 summarises how reforms of the US National Flood Insurance Program strived to better align incentives with risks. However, the distributional consequences of these reforms meant that the package was not politically viable.

OECD countries' approaches to risk transfer are diverse in their coverage, scope and financial responsibility. While some countries take a fully privatised approach, others complement compulsory coverage of natural hazards in private insurance with state support for specific risks (Keskitalo, Vulturius and Scholten, 2013). A possible limitation of a fully privatised approach is that insurance coverage becomes unaffordable to the most vulnerable

Box 5.2. **Reform of the US National Flood Insurance Program**

The US National Flood Insurance Program (NFIP) was created in 1968 to help property owners protect themselves financially against floods associated with hurricanes, tropical storms, heavy rains and other climate events. Participation in the programme is mandatory for all properties with mortgages from federally regulated or insured lenders located in high flood-risk areas (defined as a 1 in 4 chance of flooding during a 30-year mortgage). Participation is not required for properties in moderate- to low-risk areas, but they account for nearly 25% of flood insurance claims and one-third of Federal Disaster Assistance for flooding.

A 2010 report from the Government Accountability Office identified a number of design features that constrain the fiscal soundness of the programme and impede the efficient management of flood risk. The features include: statutory limits on rate increases, the inability to reject high-risk applicants, the mismatch between NFIP premiums and the real flood risks for almost a quarter of property owners, the use of "grandfathered" rates for some properties not taking into account reassessments of flood risk, and the inability of the programme to deny coverage to repetitive loss properties. These properties account for 25-30% of insurance claims while only making up for 1% of policies. A series of disasters including hurricanes Sandy and Katrina, have contributed to a program deficit of close to USD 24 billion.

To address these challenges, US Congress passed in 2012 the Biggert-Waters Flood Insurance Reform Act (BW-12). The objective of the reform was to target the fiscal soundness of the programme, promote more efficient risk management, and to assess future changes to flood risk on the basis of the best available scientific evidence. This resulted in annual premium rate increases for policyholders of up to 20% (twice the previous limit) based on calculations of an "average historical loss year", including catastrophic loss years. Subsidies were also phased out for a number of properties, severe repetitive loss properties in particular. Furthermore, flood insurance rate maps were updated to reflect updated information on changes in sea levels, precipitation and hurricane intensity.

Political opposition to BW-12 ultimately led to the passage of the Homeowner Flood Insurance Affordability Act in March 2014, which repeals and modifies certain provisions of BW-12. For example, instead of an immediate premium increase to full-risk rates, the Affordability Act requires (with a few exceptions*) that the increase is gradual, but no less than 5% and no more than 18% annually. Further, the Act reinstates the use of "grandfathered" rates that enable policyholders of new properties to benefit the first year from premium rates offered to properties located outside the Special Hazard Area. Policyholders in high-risk areas required to pay their full-risk rate under BW-12 are also entitled to refunds, while policyholders that face an 18% premium increase may be entitled to refunds. Despite these amendments, the objective of the Affordability Act is to make the NFIP self-sufficient by gradually moving towards actuarial rates. All policies for primary residences will also be subjected to a USD 25 surcharge while all other policies include a USD 250 surcharge.

* Exceptions include older business properties, older non-primary residences, severe repetitive loss properties, and buildings that have been substantially damaged or improved built before the adoption of a local Flood Insurance Rate Map that will face up to a 25% annual increase.

Source: GOA (US Government Accountability Office) (2010), *National Flood Insurance Program. Continued Actions Needed to Address Financial and Operational Issues*, Testimony Before the Subcommittee on Housing and Community Opportunity, Committee on Financial Services, House of Representatives, Washington, DC.; FEMA (2014), *Homeowner Flood Insurance Affordability Act*, US Federal Emergency Management Agency, Washington, DC.; OECD (2013b): *Water and Climate Change Adaptation: Policies to Navigate Uncharted Waters*, OECD Studies on Water, OECD Publishing, Paris, *http://dx.doi.org/10.1787/22245081*.

households. At the same time, if disaster relief by the state is extensive and unconditional, this may create a disincentive for households to purchase insurance coverage or to invest in risk reduction measures (Keskitalo, Vulturius and Scholten, 2013).

Risks held by governments can also be transferred to third parties. These mechanisms are particularly relevant to developing countries, but there are some examples of their use by OECD countries. The EU Solidarity Fund (EUSF) is a mechanism for sharing risks between EU member states. These countries can request financial support to cover non-insurable damages following a major disaster, such as the cost of restoring public infrastructure, the cost of emergency relief and clean up, and the cost of protecting cultural heritage (Bräuninger et al., 2011). For example, in 2002 Austria received EUR 134 million from the EUSF after being affected by EUR 2.9 billion of total flood losses (Bräuninger et al., 2011). Climate change is projected to increase the value of claims on the EUSF arising from flooding (Jongman et al., 2014).

Reinsurance and capital markets can provide alternative sources of risk-bearing capacity. Reinsurance companies operate globally and play an important role in diversifying risks across national borders and ensuring access to disaster risk capital. Similarly, catastrophe bonds can be used to transfer risks to capital markets. Purchasers of catastrophe bonds receive an attractive return if a pre-defined event (e.g. measured by a parametric threshold such as wind speed or rainfall) does not occur. However, if the threshold is passed, the investor will lose their investment and the money is used to cover the costs of the catastrophe (Suarez and Linnerooth-Bayer, 2011). The implementation of catastrophe bonds by national governments has been hindered by the cost and complexity of issuance. One exception, however, is Mexico's USD 315 million catastrophe bond launched in 2012. The bond provides coverage against earthquakes and hurricanes, building on the success of the 2009 MultiCat transaction (OECD, 2013f).

Absorbing impacts arising from residual risks

Governments face potential liabilities arising from their exposure to the risks from climate change that have not been reduced or transferred to others. The combination of climate change and socio-economic changes are likely to increase the scale of these liabilities (IPCC, 2014a). Many of these liabilities are contingent in the sense that they may only be realised if an extreme event occurs, or a slow onset event passes a certain threshold (e.g., infrastructure being submerged due to rising sea levels). Extreme events present a particular challenge, as the scale and likelihood of events cannot be reliably calculated. Thus, the true extent, or even the existence, of a liability may only become apparent after an event occurs (OECD, 2013e). Efforts to improve understanding, disclosure and management of these financial liabilities are essential elements of governments' preparations for the effects of climate change.

Governments are exposed to a wide range of direct fiscal pressures arising from climate risks. Common practice in OECD countries is for national governments to self-insure their own operations, which means that they must fund the replacement of assets damaged by extreme events (Warner et al., 2012). In addition to this, national budgets may have to absorb costs resulting from commitments to compensate losses affecting households, businesses, or other tiers of government. These commitments include both the ex-ante risk transfer mechanisms discussed above, as well as measures that are put in place following an event: for example, temporary tax relief for areas affected or compensation payments arising from the expectation that governments will step in as an "insurer of last resort" to cover uninsured losses (OECD, 2013e).

In addition to these financial outlays that can be directly traced to the effects of climate change, governments may also have significant indirect liabilities. Climate change will have macroeconomic effects, including shifts in the terms of trade, impacts on productivity, disruption from extreme weather events and changes in patterns of tourism (Vivid Economics, 2013). These macroeconomic effects will affect government finances through channels including tax revenues and impacts on social payments. In principle, these could serve to offset or amplify the direct impacts of climate change. Indirect impacts of climate change are not routinely tracked and there is limited empirical data on their likely importance. Some indicative evidence (provided in Box 5.3) suggests that indirect costs are likely to be significant and additional to direct costs.

Box 5.3. **Climate change impacts and adaptation needs in Germany, Finland, and Italy**

A study undertaken by the Centre for European Policy Studies examines climate impacts and adaptation needs in Germany, Finland, and Italy. It finds that direct costs caused by climate change in Germany could be as high as EUR 3.4-15.9 billion per year by the end of the century. Indirect costs measured in terms of lost tax revenue could amount to EUR 22.9-104.6 billion per year driven by reductions in consumer income or investments in less productive capital. The Finish study identified the tourism sector, followed by the forestry and agriculture sectors, as the sectors most affected financially by the impacts of climate change albeit in a positive direction. Energy and hydrology are projected to face negative consequences. However, the overall fiscal effects for Finland are not expected to be severe compared to other European countries because of the balance between positive and negative effects. In Italy, sea level rise is projected to contribute to fiscal pressures from the required large-scale public investment representing the majority of overall adaptation investment. The indirect effects in Finland and Italy are not estimated in this analysis.

Source: CEPS (Centre for European Policy Studies) (2010), "The Fiscal Implications of Climate Change Adaptation", *Final Report*, No. ECFIN/E/2008/008, Mannheim.

Governments have a range of options for absorbing the liabilities arising from residual risks. Each option is characterised by costs to the government and by factors that constrain availability and suitability (Mechler et al., 2010; G20/OECD, 2012; OECD, 2013e):

- **Contingency funds or reserves:** Ring-fenced government reserves can help to absorb the consequences of extreme events. The size of these funds represents a trade-off between the opportunity costs of holding reserves, against the risk that the fund will be insufficient to cover losses.
- **Budget diversion:** Governments can reallocate funds earmarked for other purposes to cover the costs of an extreme weather event. Forced reallocation of resources can be disruptive and will forgo the benefits of the intended investment.
- **Taxes:** After an extreme weather event, the government may introduce new taxes or raise existing taxes to cover additional expenditures.
- **Borrowing:** Governments can also borrow money from the domestic private sector or internationally by issuing government bonds.

Each of these options brings with it a set of trade-offs. OECD countries have had considerable flexibility to date in how they finance impacts using this range of tools. However, the rising scale of potential losses from climate change, in addition to continuing pressures on government finances, means that this could become more challenging in future.

Advance preparation for these events can help governments design a strategy appropriate to their circumstances and put in place any necessary ex-ante arrangements (G20/OECD, 2012). At the national level, this process of identifying liabilities could entail an assessment of the vulnerability of public-sector assets, an overview of existing (legal or implicit) commitments to cover private sector losses, and an understanding of the indirect impacts resulting from climate-related impacts on economic activity. This can be complemented with an assessment of overall risk financing (e.g. reserves and credit) and risk transfer (e.g. insurance, catastrophe bonds or reinsurance).

Good practice is for potential contingent liabilities, whether related to climate change or not, to be identified and disclosed as a statement of risks alongside the budget (Lindwall, 2013). In Australia, for example, as part of the annual budget process the government discloses contingent liabilities based on two criteria: i) that the potential impact of the contingent liabilities is higher than AUD 20 million in any one year, or ii) higher than AUD 50 million over the forward estimated period. In principle, these could include liabilities relating to risks arising from climate change, but in practice these potential liabilities are not explicitly listed.

Financing adaptation at the sub-national level

National measures to finance adaptation will need to be complemented by measures at the sub-national level. Local governments will not be responsible for financing all adaptation measures, but like national governments, they have an important role in establishing a regulatory framework that encourages climate resilient investments and growth. At the sub-national level, the need for adaptation financing is often focused on the urban, rather than the rural level (IPCC, 2014a). This may in part be explained by the relatively high concentration of essential assets prone to climate change in urban areas such as schools, hospitals, water supplies, communications, infrastructure and roads (IPCC, 2014a).

The high concentration of assets in cities is complemented by high exposure to climate hazards since cities traditionally have been located near rivers and oceans to facilitate easy access and connectivity to other cities (World Bank, 2010a). In Europe, an estimated 70% of the largest cities are vulnerable to sea level rise, the majority of which are located less than 10 metres above sea levels (World Bank, 2010a). Similarly, Miami, New York City and Tokyo are among the top 20 cities in the world where both people and assets are particularly exposed to coastal flooding (IPCC, 2014). However, with rising sea levels and increased frequency and intensity of weather events, this geographic advantage is affecting cities' vulnerability to climate change.

At the global level, the World Bank (2010b) estimates that 80% of investments in risk reduction for climate change will be in sectors related to urban areas. Climate change is only one of many issues demanding the attention of urban policy planners. However, the established practice of local governments to respond to a wide variety of risks provides an important foundation for addressing climate change (IPCC, 2014a). In OECD countries, for example, local governments already finance around 70% of public investment and 50% of public spending on environmental protection (e.g. waste and water management and the protection of biodiversity and landscapes) (OECD, 2010). Potential sources of local finance include revenues generated from service payments, taxes or user fees on public resources. Similarly, through the application of revolving funds, potential costs saved from energy efficient investments in municipal buildings can be allocated for investments in adaptation (IPCC, 2014a). Public-Private Partnerships (PPPs) and other market-based finance instruments such as green bonds could also provide an important source of financing for adaptation at the local level.

Some characteristics of urban adaptation may justify financial contributions from national governments. These include the potential spill-over benefits of local adaptation measures to the resilience of the country as a whole. Financial contributions to local adaptation may also be made on the basis of solidarity or other distributional considerations. This can for example be in terms of grants, loans, bonds or by transferring other sources of revenue collected at the national or regional level to local government (IPCC, 2014a). In practice, however, support for urban adaptation remains uneven. A survey of 468 cities in both developed and developing countries found that 60% of the cities surveyed did not receive any support for their adaptation initiatives (Carmin et al., 2012). Table 5.3 summarises the main sources and financial instruments that local governments can draw upon to fund adaptation.

Table 5.3. **Main sources of funding and financial instruments for urban adaptation**

Sources of funding	Types	Instruments	What can be funded (with some examples of funds)	Urban capacity required to access funding
Local: public	Local revenue raising policies; taxes, fees, and changes or use of local bond markets	• Local taxes (e.g. on property, land value capture, sakes, business, personal, income, vehicles) • User charges (e.g. water, sewers, public transport, refuse collection) • Other charges or fees (e.g. parking, licenses)	• Urban infrastructure and services • Urban adaptation programmes and planning processes • Urban capacity building	Cities with well-functioning administrative and institutional capacity and adequate funding from local revenue generation and intergovernmental transfers
Local: public-private	Public-Private Partnerships (PPP) contracts and concessions	• Concessions and private finance initiatives to build, operate and/or maintain key infrastructure • Energy performance contracting • Municipal bonds	• Medium to large-scale infrastructure with strong private goods (to allow rents for private sector)	Cities with strong capacity for legal oversight and management
Local and national: private-public	National and local financial markets	• Commercial loans • Private bonds • Municipal bonds	• Basic physical infrastructure (need for collateral)	Well-functioning local or national financial markets that city governments can access
National: public	National (or state/provincial) revenue transfers or incentive mechanisms	• Revenue transfers from central or regional government • Payment for ecosystem services or other incentive measures	• Urban payment for environmental services in Brazil • Sweden's KLIMP climate investment programme	Cities with good relations with national governments, strong administrative capacity to design and implement policies and plans
International: private	Market-based investment	• Foreign direct investment, joint ventures	• Industrial infrastructure • Power generation infrastructure	Cities with strong national enabling conditions and policies for investments
International sources	Grants, concessional financing (e.g. Adaptation Fund)	• Grants, concessional loans, and loan guarantees through bilateral and multilateral development assistance • Philanthropic grants	• Urban capacity building • Urban infrastructure adaptation planning	Strong multi-level governance – cities with good relations with national governments.

Source: IPCC (2014a), *Climate Change 2014: Impacts, Adaptation, and Vulnerability*, Contribution of Working Group II to the Fifth Assessment Report of the Intergovernmental Panel on Climate Change, Field, C.B. et al. (eds.), Cambridge University Press, Cambridge, United Kingdom and New York, NY, USA.

In rural areas, finance for risk reduction is primarily allocated through sectoral mechanisms, particularly for the water, agriculture and the forestry sectors. A survey of EU member states found that project-based public support was the most common financing mechanism in place for implementing adaptation in sectors identified by member states as relevant to adaptation. This was followed by explicit budget allocations (EEA, 2014). The extensive use of project-based financing is somewhat surprising given the push for countries to take an integrated approach to adaptation planning and implementation.

Financing adaptation at the sectoral level

The main financing needs for risk reduction will arise at the sectoral level. Across sectors, a number of measures have been identified that are effective in reducing the exposure and vulnerability to climate risk, while at the same time being implementable at relatively low costs (OECD, 2008). For example, "soft" adaptation measures such as farm-level adjustments (e.g. weather proofing barns) may entail low upfront costs but can lead to significant benefits in terms of offsetting damages. This is also the case for other behavioural adaptation measures, such as enhanced water use efficiency.

The diverse nature of the individual sectors (e.g. in terms of their business structures and their roles within the economy) means that the most suitable approach to financing adaptation will vary greatly. Such measures may be matched by public support for private risk transfer mechanisms, administrative cost-sharing schemes, or public insurance of last resort. Across all sectors, it is unlikely that individual mechanisms can address the financing challenge. Instead a portfolio of approaches may be needed to generate the estimated financial flows needed.

There are, however, generic barriers that need to be overcome in order to improve the effectiveness and efficiency of sectoral financing for adaptation. The following barriers identified in the context of energy efficiency are also relevant for the context of adaptation and include: i) direct financial barriers (e.g. payback expectations, investment horizons, competing purchase decisions, and price signals), and ii) institutional and administrative barriers (e.g. informational barriers, professional skills and knowledge, and awareness of potential benefits) (BPIE, 2011). Some financing tools currently being applied in the: i) agriculture, ii) energy, transport and building, and iii) water sectors are summarised in Table 5.4. A common feature of all three sectors is their long history of adjusting practices to emerging risks.

Agriculture is a key sector for adaptation, given its sensitivity to climate. Historically, assistance to the agricultural sector has been provided by price support schemes and budgetary payments to farmers. However, distorting production and input assistance measures risk weakening farmers' incentives to identify more resource-efficient and sustainable processes (OECD, 2014b). Moreover, this type of government assistance can serve to offset "normal" adjustment pressures from the market, impeding ongoing structural change and preventing more efficient farmers from expanding their operations. In that context, it is encouraging that the total volume of support (relative to farm incomes) has been declining in OECD countries. In addition, the composition of that support has shifted with a reduction in the share of the potentially most distorting types of support (OECD, 2014b).

The EU Common Agricultural Policy (CAP) is an example of such reforms aiming to incorporate environmental objectives into funding allocations, explicitly encouraging the financing of climate-friendly practices. A CAP policy instrument, the Green Direct Payment,

Table 5.4. **Examples of financing for adaptation at the sectoral level**

	Tools to reduce risks	Tools to transfer risks
Agriculture	• Increasing investments in developing climate resilient crops • Making access to funding contingent on climate-friendly practices (e.g. recent reforms of the EU Common Agricultural Policy) • Education and training to make farmers better aware of the risks and the risk reduction measures available • Removing market distortions (e.g. by pricing water usage appropriately)	• Transitional income provided to farmers experiencing economic hardship (e.g. the Farm Household Allowance in Australia) • Public/private crop insurance schemes to e.g. cover yield losses and inventory losses from extreme weather events
Energy, transport and building	• Updating infrastructure codes and standards to take adaptation into account • Investments in retrofitting existing infrastructure to make it more climate resilient • Reporting obligations of key utilities makes the operators aware of the risks and provides the government with a strategic overview of vulnerabilities of key infrastructure (e.g. the UK Adaptation Reporting Power)	• Public-private partnerships can split the risks inherent in any project developments between the public and the private sector
Water	• Tradable water permits and water pricing • Regulatory incentives to enhance water efficiency water recycling and water management (e.g. Australia's Water for the Future programme and Mexico's 2030 Water Agenda) • Investments in flood forecasting, flood protection and coastal erosion management	• Weather insurance

rewards farmers for respecting three obligatory agricultural practices: i) maintenance of permanent grassland, ii) ecological focus areas, and iii) crop diversification on arable land. In Australia, the Management Deposit Scheme enables farmers to set aside pre-tax deposits from primary production income during high-earning years that can be withdrawn as (taxed) income in years of low income (Australian National Rural Advisory Council, 2012). Similarly, the Farm Household Allowance provides transitional income support to farmers that experience economic hardship, regardless of the cause, for up to three years in addition to training and consultancy services (Australian Department of Agriculture, 2014).

Across the OECD, infrastructure investments lag behind the financing needed for sectors such as energy, transport and building. Climate change will increase the scale of this challenge: new infrastructure will have to be built that meets current resilience standards at the same time as existing infrastructure will have to be retrofitted to meet those standards. For example, it has been projected that global investments in new physical assets will triple between 2000 and 2030 (UNFCCC, 2007). Many of these investments have a lifetime of 30 years or more, and will to a large extent (86%) be financed by the private sector (UNFCCC, 2007). National governments can play an essential role in ensuring that investments in new capital formation are climate resilient and do not increase the country's vulnerability to climate change by setting clear, long-term policy goals (Corfee-Morlot et al., 2012).

The design of PPPs can help to ensure that the right incentives are in place. Compared to traditional public infrastructure investments, under PPPs, part of the risk inherent in the project is handed over to the private agent (Irwin and Mokdad, 2010). The private sector agent has an incentive to decrease those risks by performing well so that investment returns are secured (Corfee-Morlot et al., 2012). However, the effectiveness of this depends crucially upon the detail of the relevant contracts, as well as the signatories' ability to honour their commitments. Climate can give rise to additional risks during the construction and operation of the facilities that may not have been foreseen at the signing of the contract. The allocation of those risks will affect the incentives to manage them. Interdependencies can mean that a lack of robustness in a single infrastructure network can impose economy-wide costs. As such, it is essential that the relevant risks are identified in advance and responsibility for managing those risks is clearly identified.

Only a few countries have begun to address the issue of financing adaptation for water systems (OECD, 2013a). Countries that have started to address financing issues for water have used a range of approaches. For example, Australia, Canada, France and Sweden have dedicated general adaptation funding from public budgets at the national level to the water sector. By contrast, adaptation support for water-related projects in the Netherlands and the Czech Republic is channelled through the Delta Fund and the Flood Prevention Programme respectively (OECD, 2013b). A few OECD countries (e.g. Chile, Estonia, Hungary, Mexico, Slovenia and Turkey) have received funding from international funding mechanisms (including EU Structural Funds and Cohesion Fund) to advance adaptation of water systems.

Several OECD countries are also exploring potential new sources of financing and innovative mechanisms within the water sector. Programmes include water trading, efficient water pricing, incentives for ecosystems-based adaptation and green infrastructure. Tradable permits provide a mechanism for managing transitions in the availability of water. Many systems are administered by the government and as such, provide the government with leverage within the system and revenues from permit rents. There are hundreds of water transfers in California each year; the majority of which are transfers between agricultural users within the same water basin, some in the form of tradable permits. Water transfers are used to help meet in-stream demands from government agencies, such as the state's Environmental Water Account. In officially declared emergency situations (e.g. drought) the California Department of Water Resources opens a California Drought Water Bank, which buys surplus water allocations from northern California and sells (and transports) these allocations to areas in southern California hard-hit by drought.

Adaptation for biodiversity and ecosystems has the potential to reduce the consequences of climate change for those systems, but also reduce climate risks faced by communities. Possible measures include: watershed management to protect against droughts and floods; rangeland management to prevent desertification; sustainable management of fisheries and forests to ensure food security; and mangrove restoration to buffer against storm surges.

OECD (2013d) explores innovative financing tools for biodiversity that can be related to activities on adaptation for biodiversity and ecosystems. Those tools include: environmental fiscal reform; payments for ecosystem services; biodiversity offsets; markets for green products; biodiversity in climate change funding; and biodiversity in international development finance. Governments need to consider carefully the drivers of biodiversity loss; the governance and institutional capacity needed; and socio-economic, cultural and political circumstances to make sure that those financing tools are effective (OECD, 2013d). The United States and Australia have introduced adaptation financing tools for biodiversity and ecosystems, which are summarised in Box 5.4.

Box 5.4. **Adaptation financing for biodiversity and ecosystems**

As with agriculture, biodiversity and ecosystems will both be affected by climate change, while also having a broader systemic importance. Existing conservation frameworks can contribute to the adaptation of ecosystems faced with stresses from climate change. However, new mechanisms can contribute to the efficient support of biodiversity in a changing climate, particularly in light of non-climate stressors. One recent development in caring for biodiversity and ecosystems is the use of habitat banking* to raise funds for adaptation measures (e.g. forest seed banks) and to encourage diverse management practices on land rich in biodiversity.

In the United States, the National Fish, Wildlife, and Plants Climate Adaptation Strategy is used to guide ecosystem adaptation and resilience efforts. One of its goals is to reduce non-climate stressors so that ecosystems are more resilient to climate change. Furthermore it aims to support adaptive habitat and wildlife management that can evolve with a changing climate and to incorporate climate change knowledge into better management practice to secure valuable resources. Implementation of the strategy is in progress. The value of the natural resource is translated into quantified "credits" that usually include long-term funding for the management and protection of the natural resource.

In New South Wales, Australia, BioBanking provides a mechanism for linking developers who wish to offset negative environmental impacts (amongst others) with landowners who wish to involve and manage their land for biodiversity conservation and adaptation purposes. The landowner receives funds to maintain and improve biodiversity value. To ensure consistency in biodiversity values developers are obliged to buy credits for "the same vegetation type or another vegetation type in the same formation that contain the same predicted species" or "a more cleared vegetation type that contains the same threatened species". The effect of this is to increase the flows of private resources for habitat conservation, to complement existing public sector resources.

* Habitat banking is a biodiversity compensation mechanism that is based on the concept of biodiversity offsets which are, according to BBOP (2009): "measurable conservation outcomes resulting from actions designed to compensate for significant residual adverse biodiversity impacts arising from project development and persisting after appropriate prevention and mitigation measures have been implemented. The goal of biodiversity offsets is to achieve no net loss, or preferably a net gain, of biodiversity on the ground with respect to species composition, habitat structure and ecosystem services, including livelihood aspects".

Source: National Fish, Wildlife and Plants Climate Adaptation Paternership (2012), *National fish, Wildlife and Plants Climate Adaptation Strategy*, Association of Fish and Wildlife Agencies, Council on Environmental Quality, Great Lakes Indian Fish and Wildlife Commission, National Oceanic and Atmospheric Administration, and US Fish and Wildlife Service, Washington, DC; McKenney and Kiesecker, 2010; Environment & Heritage, 2014; Department of Environment, Climate Change and Water New South Wales (2009), *The Science Behind BioBanking*, Australian Government, South Sydney.

References

Australian Department of Agriculture (2014), "Farm Household Allowance Questions and Answers", Australian Government, *www.agriculture.gov.au/agriculture-food/drought/assistance/income-support-for-farmers/farm-household-allowance* (accessed 26 November 2014).

Australian Government (2011), "Budget Strategy and Outlook 2011-12: Statement 8 – Statement of Risks", Australian Government, *www.budget.gov.au/2011-12/content/bp1/html/bp1_bst8.htm* (accessed 26 November 2014).

Australian National Rural Advisory Council (2012), *Report on the Effectiveness of the Farm Management Deposit Scheme*, National Rural Advisory Council, Canberra.

Binder, S.B. (2013), *Resilience and Post-disaster Relocation: A Study of New York's Home Buyout Plan in the Wake of Hurricane Sandy*, Report prepared for Quick Response Grant Program, Natural Hazards Center, University of Colorado.

BPIE (2011), *Europe's buildings under the microscope – A country-by-country review of the energy performance of buildings*, Building Performance Institute Europe, Brussels.

Bräuninger, M. et al. (2011), *Application of economic instruments for adaptation to climate change*, Perspectives GmbH, Hamburg.

Carmin, J., N. Nadkarni and C. Rhie (2012), *Progress and Challenges in Urban Climate Adaptation Planning: Results of a Global Survey*, MIT, Cambridge.

CEPS (2010), "The Fiscal Implications of Climate Change Adaptation", *Final Report*, No. ECFIN/E/2008/008, Centre for European Policy Studies, Mannheim.

Corfee-Morlot, J. et al. (2012), "Towards a Green Investment Policy Framework: The Case of Low-Carbon, Climate-Resilient Infrastructure", *OECD Environment Working Papers*, No. 48, OECD Publishing, Paris, *http://dx.doi.org/10.1787/5k8zth7s6s6d-en*.

Dellink, R. et al. (2014), "Consequences of Climate Change Damages for Economic Growth: A Dynamic Quantitative Assessment", *OECD Economics Department Working Papers*, No. 1135, OECD Publishing, Paris, *http://dx.doi.org/10.1787/5jz2bxb8kmf3-en*.

Department of Environment, Climate Change and Water New South Wales (2009), *The Science Behind BioBanking*, Australian Government, South Sydney.

EEA (2014), *National adaptation policy processes in European countries: 2014*, European Environment Agency, Copenhagen.

EEA (2007), "Climate change: The cost of inaction and the cost of adaptation", *EEA Technical Report*, No. 13/2007, European Environment Agency, Copenhagen.

Environment & Heritage (2014), *Biobanking*, *www.environment.nsw.gov.au/biobanking/* (accessed 11 August 2014).

EC (2014), "Commission Implementing Decision of 19 March 2014 on the adoption of the LIFE multiannual work programme for 2014-17", *Official Journal of the European Union*, No. L 116/1, European Commission, Brussels.

Elrick-Barr, C.E. et al. (2014), "Toward a new conceptualization of household adaptive capacity to climate change: applying a risk governance lens", *Ecology and Society*, Vol. 19(4) *http://dx.doi.org/10.5751/ES-06745-190412*.

FEMA (2014), *Homeowner Flood Insurance Affordability Act*, US Federal Emergency Management Agency, Washington, DC.

GAO (US Government Accountability Office) (2010), "National Flood Insurance Program. Continued Actions Needed to Address Financial and Operational Issues", Testimony Before the Subcommittee on Housing and Community Opportunity, Committee on Financial Services, House of Representatives, Washington, DC.

GAO (2005), *Catastrophe Risk: US and European Approaches to Insure Natural Catastrophe and Terrorism Risks*, Report to the Chairman, Committee on Financial Services, House of Representatives, Washington, DC.

Hansen, R.E. (2012), "Climate Change Disclosure by SEC Registrants: Revisiting the SEC's 2010 Interpretive Release", *Brooklyn Journal of Corporate, Financial and Commercial Law*, Vol. 6/2.HM Government (2014), *Water Act gains Royal Assent*, *www.gov.uk/government/news/water-act-gains-royal-assent*, HM Government (accessed 12 August 2014).

HM Government (2013), *Flood insurance agreement reached*, *www.gov.uk/government/news/flood-insurance-agreement-reached*, HM Government (accessed 12 August 2014).

IEG (2013), *Adapting to Climate Change: Assessing the World Bank Group Experience – Phase III*, Independent Evaluation Group, the World Bank Group, Washington, DC.

IPCC (Intergovernmental Panel on Climate Change) (2014), *Climate Change 2014: Impacts, Adaptation, and Vulnerability*, Contribution of Working Group II to the Fifth Assessment Report of the Intergovernmental Panel on Climate Change, Cambridge University Press, Cambridge and New York.

IPCC (2013), *Climate Change 2013: The Physical Science Basis*, Contribution of Working Group I to the Fifth Assessment Report of the Intergovernmental Panel on Climate Change, Cambridge University Press, Cambridge and New York.

Irwin, T. and T. Mokdad (2010), *Managing Contingent Liabilities in Public-Private Partnerships: Practice in Australia, Chile, and South Africa*, World Bank and the Public-Private Infrastructure Advisory Facility, Washington, DC.

Israel, N.D. (2013), *Inaction on Climate Change: The Cost to Taxpayers*, Ceres, Boston, MA.

Jongman, B. et al. (2014), "Increasing Stress on Disaster-Risk Finance due to Large Floods", *Nature Climate Change*, Vol. 4, *http://dx.doi.org/10.1038/nclimate2124.*

Keskitalo, E.C.H., G. Vulturius and P. Scholten (2013), "Adaptation to Climate Change in the Insurance Sector: Examples from the UK, Germany and the Netherlands", *Natural Hazards*, Vol. 71(1), *http://dx.doi.org/10.1007/s11069-013-092-7.*

Lindwall, P. (2013), *Budgeting for Contingent Liabilities Discussion Paper,* OECD, Paris, [GOV/PGC/SBO(2013)7], *www.oecd.org/officialdocuments/publicdisplaydocumentpdf/?cote=GOV/PGC/SBO%282013%297&doclanguage=en* (accessed on 15 April 2015).

McFarland, J.M. (2008), "Warming Up to Climate Change Risk Disclosure", *Fordham Journal of Corporate & Financial Law*, Vol. 14(2).

McKenney, B.A. and J.M. Kiesecker (2010), "Policy Development for Biodiversity Offsets: A Review of Offset Frameworks", *Environmental Management*, Vol. 45(1), *http://dx.doi.org/10.1007/s00267-009-9396-3.*

Mechler, R. et al. (2010), "Assessing the Financial Vulnerability to Climate-Related Natural Hazards", *World Bank Policy Research Working Paper Series*, No. 5232, World Bank, Washington, DC.

Mullan, M., et al. (2013), "National Adaptation Planning: Lessons from OECD Countries", *OECD Environment Working Papers*, No. 54, OECD Publishing, Paris, *http://dx.doi.org/10.1787/5k483jpfpsq1-en.*

National Fish, Wildlife and Plants Climate Adaptation Paternership (2012), *National fish, Wildlife and Plants Climate Adaptation Strategy*, Association of Fish and Wildlife Agencies, Council on Environmental Quality, Great Lakes Indian Fish and Wildlife Commission, National Oceanic and Atmospheric Administration, and US Fish and Wildlife Service, Washington, DC.

OECD (2014a), *Boosting Resilience through Innovative Risk Governance*, OECD Publishing, Paris, *http://dx.doi.org/10.1787/9789264209114-en.*

OECD (2014b), *Agricultural Policy Monitoring and Evaluation 2014: OECD Countries*, OECD Publishing, Paris, *http://dx.doi.org/10.1787/agr_pol-2014-en.*

OECD (2013a), *Water and Climate Change Adaptation: Policies to Navigate Uncharted Waters*, OECD Studies on Water, OECD Publishing, Paris, *http://dx.doi.org/10.1787/9789264200449-en.*

OECD (2013b), *Water and Climate Change Adaptation: An OECD Perspective*, OECD Studies on Water, OECD Publishing, Paris, *http://dx.doi.org/10.1787/9789264200449-en.*

OECD (2013c), *OECD Environmental Performance Reviews: Austria 2013*, OECD Publishing, Paris, *http://dx.doi.org/10.1787/9789264202924-en.*

OECD (2013d), *Scaling-up Finance Mechanisms for Biodiversity*, OECD Publishing, Paris, *http://dx.doi.org/10.1787/9789264193833-en.*

OECD (2013e), "The Case for Managing Risks Related Contingent Liabilities in Public Finance Frameworks", report prepared for the High Level Risk Forum, 12-13 December, OECD, Paris.

OECD (2013f), *OECD Reviews of Risk Management Policies: Mexico 2013: Review of the Mexican National Civil Protection System*, OECD Publishing, Paris, *http://dx.doi.org/10.1787/9789264192294-en.*

OECD (2010), *Cities and Climate Change*, OECD Publishing, Paris, *http://dx.doi.org/10.1787/9789264091375-en.*

OECD (2008), *Economic Aspects of Adaptation to Climate Change: Costs, Benefits, and Policy Instruments*, OECD Publishing, Paris, *http://dx.doi.org/10.1787/9789264046214-en.*

OECD/G20 (2012), *Disaster Risk Assessment and Risk Financing. A G20/OECD Methodological Framework*, OECD, Paris, *www.oecd.org/gov/risk/G20disasterriskmanagement.pdf.*

Porrini, D. and R. Schwarze (2012), "Insurance Models and European Climate Change Policies: an Assessment", *European Journal of Law and Economics*, Vol. 38, *http://dx.doi.org/10.1007/s10657-012-9376-6.*

Surminski, S. (2010), *Adapting to Extreme Weather Impacts of Climate Change – How can the Insurance Industry Help?*, Climate Wise, London.

Swiss Re (2014), "Natural catastrophes and man-made disasters in 2013: large losses from floods and hail; Haiyan hits the Philippines", *Sigma*, No. 1/2014, Swiss Re Ltd., Zurich.

Tol, R.S.J., S. Fankhauser and J.B. Smith (1998), "The scope for adaptation to climate change: what can we learn from the impact literature", *Global Environmental Change*, Vol. 8(2), *http://dx.doi.org/10.1016/S0959-3780(98)00004-1.*

UNFCCC (2013), *Opportunities for Parties included in Annex I to the Convention whose special circumstances are recognized by the Conference of the Parties to benefit from support from relevant bodies and institutions to enhance mitigation, adaptation, technology, capacity-building and access to finance*, Technical Paper, United Nations Framework Convention on Climate Change, Bonn.

UNFCCC (2007), *Investment and Financial Flows to Address Climate Change*, UNFCCC, Bonn.

UNISDR (2014), *Progress and Challenges in Disaster Risk Reduction: A contribution towards the development of policy indicators for the Post-2015 Framework on Disaster Risk Reduction*, United Nations Office for Disaster Risk Reduction, Geneva.

US Fish and Wildlife Service (2014), *National Fish, Wildlife & Plants Climate Adaptation Strategy, Goals*, US Fish and Wildlife Service, *www.wildlifeadaptationstrategy.gov/goals.php* (accessed 11 August 2014).

Vivid Economics (2013), "The Macroeconomics of Climate Change", a report prepared for Defra, *www.vivideconomics.com/index.php/publications/the-macroeconomics-of-climate-change* (accessed 2 March 2015).

Warner, K. et al. (2012), "Insurance solutions in the context of climate change-related loss and damage: Needs, gaps, and roles of the Convention in addressing loss and damage," *MCII Policy Brief* No. 6, United Nations University Institute for Environment and Human Security (UNU-EHS), Bonn.

World Bank (2011), *Guide to Climate Change Adaptation in Cities*, World Bank, Washington, DC.

World Bank (2010a), "Cities and Climate Change: An Urgent Agenda", *Urban Development Series Knowledge Papers*, Vol. 10, World Bank, Washington, DC.

World Bank (2010b), *The Economic of Adaptation to Climate Change*, World Bank, Washington, DC.

Zurich Insurance Group (2014), *European floods: using lessons learned to reduce risk*, Zurich Insurance Group, Zurich.

Climate Change Risks and Adaptation
Linking Policy and Economics

Chapter 6

Tools to mainstream adaptation into decision-making processes

This chapter examines the tools that can be used to mainstream, or integrate, adaptation into existing decision making and appraisal processes. Informed by countries' experiences to date, it explores how adaptation can be included within traditional decision-making tools such as cost-benefit analysis, multi-criteria analysis and cost-effectiveness analysis. It analyses the applicability of new approaches, such as real options analysis, that are designed to support decision making under uncertainty. It discusses the importance of aligning the tools used with the institutional context that they will operate in.

Key messages

- Mainstreaming requires the integration of adaptation into existing processes and decision cycles. This recognises that adaptation will often be one of many policy objectives, and not necessarily the dominant one. It also requires the identification of suitable entry points in the policy process for introducing mainstreaming, noting these will differ across sectors and national contexts.
- To effectively mainstream adaptation, there is a need for pragmatism, capacity building and support for considering climate risks. Success will often be contingent on the timing of such mainstreaming measures and the stage at which they are considered during the decision-making process.
- With greater mainstreaming, adaptation will be further integrated into existing appraisal processes. However, adaptation involves some challenges for existing methods, due to the high uncertainty involved. A suite of new decision support tools have therefore emerged that can assist better decision-making under climate change uncertainty.
- These new methods for decision-making can be resource-intensive and complex to use. These have many potential applications for major policy initiatives and investments. To support mainstreaming, the priority is to develop "light-touch" approaches that capture the core concepts of these new methods while balancing pragmatism with economic rigour.

Adaptation mainstreaming

Adaptation mainstreaming is the integration of adaptation into decision making across a range of policy areas, rather than through the implementation of standalone adaptation measures. Mainstreaming adaptation into policy-making is a continuing process, requiring the integration of climate considerations into existing policy and project cycles. A key element of this is the integration of climate risks into the decision-support tools that are used in standard policy and project appraisal. The concept of mainstreaming – of incorporating an additional factor to the decision process – is not novel to adaptation. It has been applied across a range of policy areas to address cross-cutting issues such as gender, health and environmental impacts, and these provide useful lessons (see Box 6.1 for some examples).

Drawing on countries' experiences, this chapter discusses the key entry points and processes for mainstreaming adaptation, and how new economic tools and approaches can be applied to them. However, early experience shows that many of these approaches can be resource-intensive and complex to use. There is a continuing need to develop "light-touch" approaches that are pragmatic, fit-for-purpose and consider the likely time and resources available. These should capture the conceptual aspects of these new approaches, while maintaining a degree of economic rigour, both at policy and project level.

This chapter draws on the research, analysis and review of the ECONADAPT project, funded by the European Union's Seventh Framework Programme for research, technological development and demonstration, as well as funding provided by Canada's International Development Research Centre.[1]

Box 6.1. **Existing examples of mainstreaming**

Climate change adaptation is not the first issue to face the challenge of mainstreaming: other initiatives have been successfully mainstreamed in many OECD countries, with notable examples being health, gender and the wider environmental concerns. These provide useful lessons for climate mainstreaming. In the context of health, health impact assessments provide a set of procedures, methods and tools that can help assess the potential effect of various initiatives on the health of a population and the distribution of those effects within the population. In the case of gender, it has been emphasised that mainstreaming does not simply entail increasing the number of women in certain roles; instead, it entails changing the social consciousness so that the effects of a policy for both women and men are carefully analysed before being implemented (True, 2010). This demonstrates the ultimate goal of mainstreaming in changing perceptions and practice. Organisational arrangements for mainstreaming environmental considerations are often top-down in character, whereby the (environment) ministry pushes mainstreaming throughout the generation of guidelines that are then co-ordinated between ministries (Nunan et al., 2012).

In the context of environmental mainstreaming, the OECD (2012) study on greening development highlighted specific interventions including: using multi-year development planning processes; developing key actors' technical skills; encouraging the participation of non-government actors; building functional and technical skills; and planning and targeting efforts carefully. It also provided some recommendations on how development support can deliver better capacity by: viewing capacity support for the environment as underpinning all development support; collaborating across domestic agencies; harmonising approaches among development support providers; nurturing local ownership; focussing on results; implementing best practice guidelines; and reflecting on and learning from the process.

Source: True, J. (2010), "Mainstreaming Gender in International Relations. Gender Matters in Global Politics" (New York: Routledge) L.J. Shepherd (Ed.) pp. 189-203; Nunan, F., A. Campbell and E. Foster (2012), "Environmental Mainstreaming: The Organisational Challenges of Policy Integration", Public Administration and Development 32, 262-277, *http://dx.doi.org/10.1002/pad.1624*; OECD (2012), *Greening Development: Enhancing Capacity for Environmental Management and Governance*, OECD Publishing, Paris, *http://dx.doi.org/10.1787/9789264167896-en*.

The context for adaptation mainstreaming

The multi-sectoral characteristics of climate risks and adaptation responses have significant implications for the development of mainstreaming tools. Policy measures that will affect adaptation are often implemented for non-climate reasons, with multiple objectives and ancillary costs and benefits that are material to the overall choice of the measures. It is therefore important to understand the context for an intervention and decision, including the existing policy and objectives, non-climatic drivers, and the current decision-making process. As an example, resilience may be mainstreamed as part of an urban regeneration programme, but the design of such a programme will be dominated by local economic development objectives and other drivers, such as demographic and land-use change. Mainstreaming requires a good understanding of the individual organisations, institutional networks and processes making relevant decisions. Critically, all of these will differ with each specific adaptation context.

Another characteristic with implications for adaptation mainstreaming relates to the profile of costs and benefits over time for adaptation decisions (DFID, 2014). In many cases, the most important impacts of climate change are likely to arise in the future, say 2040 and beyond. Within economic analysis, the benefits of adapting to these changes accumulate over long time horizons, while the costs are typically incurred now. Using the public discount rates conventionally used in OECD countries, typically being between 3% and 6%, future adaptation benefits arising from climate change in the medium term and beyond are likely to be small in current terms and thus alone would rarely justify early intervention. This also has implications for the political economy of adaptation decision making, as these timescales do not always fit with political cycles or institutional time horizons. However, as noted in Chapter 3, even OECD countries have an adaptation deficit, and thus resilience to current climate variability offers a good place to start from an economic perspective.

Uncertainty about how both climate and society will change over time makes it more challenging to assess the cost-benefit profile of adaptation options (UNFCCC, 2009; Hallegatte, 2009; Wilby and Dessai, 2010). The consequences of climate change will depend on what is happening to the climate itself, but also to the societies with which it interacts. Neither of these can be reliably predicted over long time horizons. In some instances, it is unclear whether warming will lead to increased or decreased precipitation. An early adaptation response to address a long-term risk (even without discounting) has the potential to misallocate resources by over-investing in risks that do not emerge, or implementing measures that are insufficient to cope with more extreme outcomes. However, inaction as a result of uncertainties could likewise lead to much greater costs in the future, as discussed in Chapter 4.

Earlier frameworks presented adaptation responses as a set of building blocks or a spectrum of options (McGray, Hammill and Bradley, 2007; Klein and Persson, 2008). These included differentiated activities such as: addressing current vulnerability, building adaptive capacity, mainstreaming climate risks, and preparing for and tackling longer-term challenges. More recent updates of these frameworks have made them decision-led, and aligned types of activities and decisions to iterative frameworks (e.g. Ranger et al., 2010; Watkiss and Hunt, 2011). Each activity (or building block) is a different problem type, requiring different types of information, and varying methods of economic appraisal (Li, Mullan and Helgeson, 2014).

An example of such a decision-led framework is summarised in Figure 6.1. The evolution of climate change is presented at the top of the figure, as a process that starts with current climate variability and evolves over time with increasing uncertainty (Watkiss, 2014). In response, the bottom of the figure outlines three different types of adaptation response, which together address the economic and uncertainty challenges. All three types need to be considered together in an integrated adaptation strategy. The use of an adaptation pathway approach can capture and link the different activities together over time (Downing, 2012):

- First, it prioritises early actions that address the current adaptation deficit and help to build resilience for the future, particularly in the short-term when there is limited climate change apparent. This involves early capacity building and the introduction of low- and no-regret actions, which lead to immediate economic benefits (for examples, see Chapter 3). Such actions are grounded in current policy and can often use existing decision support tools.
- Second, there is early action to integrate adaptation into current decisions or activities with long life-times, such as infrastructure or planning. This requires alternative information sources and methods to the first bullet, because of the need to consider future climate change uncertainty. It also recognises that there is a need to consider options in a different way to normal appraisal, such as considering low-cost options, flexibility or robustness to address future uncertainty.

Figure 6.1. **Iterative climate risks and adaptation**

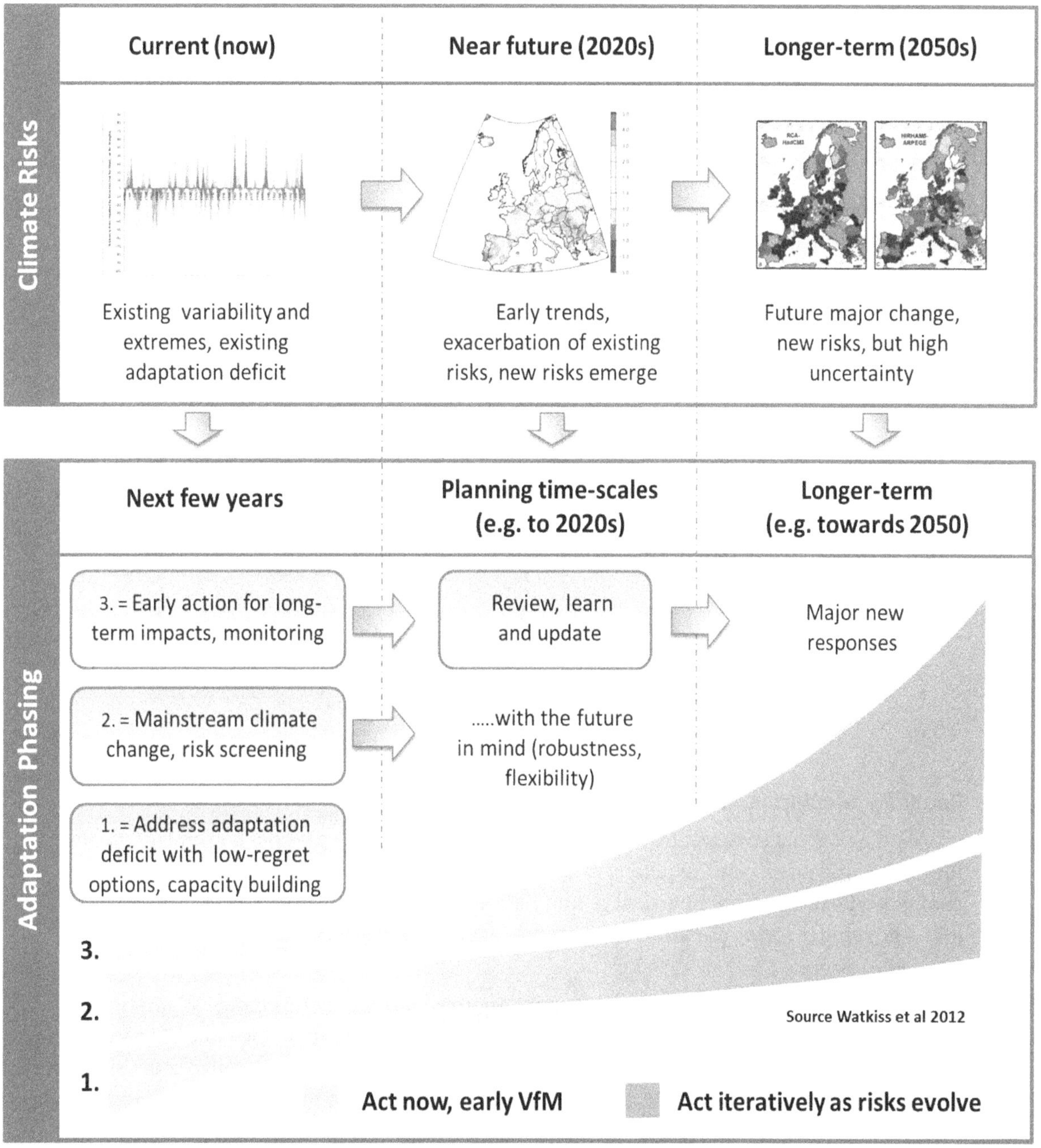

Climate Risks

Current (now)	Near future (2020s)	Longer-term (2050s)
Existing variability and extremes, existing adaptation deficit	Early trends, exacerbation of existing risks, new risks emerge	Future major change, new risks, but high uncertainty

Adaptation Phasing

Next few years	Planning time-scales (e.g. to 2020s)	Longer-term (e.g. towards 2050)
3. = Early action for long-term impacts, monitoring	Review, learn and update	Major new responses
2. = Mainstream climate change, risk screening	.....with the future in mind (robustness, flexibility)	
1. = Address adaptation deficit with low-regret options, capacity building		

3.

2.

1.

Source Watkiss et al 2012

Act now, early VfM — Act iteratively as risks evolve

Source: Watkiss (2014).

- Finally, there is a need to consider the potential major impacts of climate change, noting the possible long timescales and high uncertainty. The consideration of these longer-term issues involves important challenges and usually requires new approaches or thinking built around adaptive management. This entails learning from early activities, identifying iterative portfolios that can be brought forward or delayed according to how the future develops, and implementing early actions to address irreversibility, lock-in and encourage transformation.

Mainstreaming entry points and examples

An important component of the mainstreaming process is to find relevant entry points (OECD, 2009), that is, to identify opportunities in the national, sector or project planning process where climate risk considerations can best be integrated. This requires an understanding of the linkages between climate change adaptation and national or sector development priorities. It is also important to consider how these linkages cascade through to implementation, as well as how they are situated within the institutional and political contexts. Figure 6.2 outlines the main levels at which adaptation can be integrated into decision making at the national level. Relevant activities are shown on the left side, with decision making steps on the right. For each level, there will be an entry point for mainstreaming adaptation, i.e. the point for the initial screening and prioritisation of policies or projects. A critical part of the integration process is therefore to identify these entry points and to look for opportunities on how best to include adaptation.

Figure 6.2. **Mainstreaming steps and entry points**

Relevant activities	*Stage in Policy Cycle*	*Decision making*
	National Strategy	*Strategic level decision making e.g. creating enabling environment*
Mainstreaming in national level policies	National (Action) Planning	*Initial prioritisation of policies and programmes*
Mainstreaming in sector plans	Sector Planning	*Impact assessment and prioritisation*
Mainstreaming in sector programmes or projects	Programmes and Projects	*Detailed (economic) appraisal*

Enabling adaptation mainstreaming at the national level

At the national level, strategic decisions are taken that create the enabling environments for public- and private-sector actors, as well as communities and individuals. In the climate change context, there are now a large number of national OECD climate change strategies and a growing number of national adaptation action plans (see Mullan et al., 2013; Wilby, 2012; EEA, 2014).

In the United Kingdom, the government's approach is for climate change adaptation to be mainstreamed across all policy areas. To facilitate this process, the first UK climate change risk assessment, published in 2012, was followed up with a detailed analysis of adaptation, as part of the Economics of Climate Resilience study and the National Adaptation Programme (Frontier, 2013; HMG, 2013a; HMG, 2013b). The adaptation method for this (Watkiss and Hunt, 2011) used iterative adaptive management, drawing on the example of the Thames Estuary 2100 project (EA, 2009; 2011). It focused on mainstreaming at the sector level, working with the individual departments across government. It included a pathway analysis for a number of key risks to identify entry points and activities within existing policies and areas. The UK has also invested heavily in capacity support and in the development of tools, initially with the UK Climate Impacts Programme, and more recently with the Climate Ready team at the Environment Agency, to provide support to other policy areas for the mainstreaming approach.

Another example is the Dutch Delta programme, which has the mandate to create a strategy for the long-term protection of the coast and its hinterland (Deltacommissie, 2008).

This programme not only considers flood protection, but also includes fresh water supplies, and the wider interactions between life and work, agriculture, nature, recreation, landscape, infrastructure and energy, with a strong emphasis on sustainability. It uses an iterative adaptive management approach that prepares for the future and considers decisions in a timely fashion to plan investments (Delta Programme, 2011). It also considers short-term measures that increase adaptability (flexibility) and resistance to extreme events (robustness), to make it possible to delay reaching tipping points. Most recently, the development of adaptation plans has also been extended to consider dynamic adaptation pathways (Delta Programme, 2014; Haasnoot et al., 2013). The Netherlands has also produced a comprehensive tool for dealing with climate adaptation issues ("*handreiking ruimtelijke adaptatie*").

While recognising different circumstances between countries, lessons can be learned from the practices in developing countries. In the climate change context, developing countries have adopted a range of approaches to mainstreaming adaptation in national development strategies. For example, developing countries already include "environment" as a cross-cutting theme in their national development vision, national development plans (e.g. medium-term plans, five year plans or poverty reduction strategies), and sector development plans. In a few countries, these activities are being integrated, or at least tracked, in the national budget allocation process and in sector budget activities. Such initiatives can be extended to include climate. An example is Rwanda, which has integrated climate change (with environment) as one of seven cross-cutting issues in national development and sector development planning (Republic of Rwanda, 2012). Rwanda is also including related indicators into its budgeting process and public financial management. Some other lessons from mainstreaming practices in developing countries are heighted in Box 6.2.

Mainstreaming adaptation at programme and project levels

Existing safeguard mechanisms, such as environmental impact assessment (EIA), provide a natural entry-point for considering whether projects are vulnerable to climate change or could exacerbate climate risks elsewhere. Although originally designed to prevent negative impacts on the environment, the EIA process has the benefit of being a familiar and well-established part of the policy making process in OECD countries (Agrawala et al., 2010). It will, however, only capture those policies that are subject to environmental impact assessments, such as infrastructure construction. Moreover, it may require revision of the legal framework to include climate risks.

More generally, climate risk screening can be applied as a step in the policy-making process to identify where policies, programmes or projects may be particularly vulnerable to climate change. This has emerged strongly in relation to investment projects funded by the international finance institutions and multilateral development banks. For example, the African Development Bank (AfDB, 2011) has introduced a Climate Safeguard System that includes a traffic light system, or scorecard, to identify which projects may be particularly vulnerable to climate risk. These projects may then require a more detailed evaluation to consider integration of climate aspects into design and implementation. These evaluations tend to have a strong focus on enhancing the climate resilience of infrastructure or major investments.

A complement to the identification of high-risk policies, projects and programmes is the integration of adaptation into existing policy and project appraisal guidance. This entails the modification of existing appraisal guidance to also cover climate change or to support the consideration of some of the additional aspects and challenges of adaptation.

Box 6.2. Mainstreaming practices in developing countries

In a developing country context, mainstreaming activities usually follow a slightly different path than in developed countries, with different entry points, reflecting the differences in national strategic planning. Many developing countries are producing National Adaptation Plans (NAP). The UN guidance for the development of NAPs outlines the need for mainstreaming in developing such plans. This is critical because of the strong overlap with existing development activities (LDC Expert Group, 2012). In this context, there is a different set of entry points for mainstreaming, outlined in the table below (UNDP/UNEP, 2011) that often operate through different organisational leads. This structure closely parallels that outlined for environmental mainstreaming more generally (OECD, 2012).

Possible entry points for mainstreaming in national strategic planning policy in developing countries

Planning level	Entry point
National government and cross sector ministries	• National development vision (long-term) • Poverty reduction strategy • National development plan (e.g. 5 years) • National budget allocation process or review
Sector ministries	• Sector development plans • Sector master plans • Sector budgets
Sub-national authorities	• Decentralisation plans • District plans • Subnational budgets

Source: LDC Expert Group (2012), *National Adaptation Plans. Technical guidelines for the national adaptation plan process*, Bonn: UNFCCC secretariat. Bonn, Germany. December 2012. Available at: *http://unfccc.int/NAP*; UNDP /UNEP (2011), *Mainstreaming Climate Change Adaptation into Development Planning: A Guide for Practitioners*, UNDP-UNEP Poverty-Environment Initiative, Nairobi; OECD (2012), *Greening Development: Enhancing Capacity for Environmental Management and Governance*, OECD Publishing, Paris, *http://dx.doi.org/10.1787/9789264167896-en*.

For example, in the United Kingdom, supplementary "Green Book" guidance was published to support policy makers in accounting for adaptation in economic policy appraisal. This recommended an iterative adaptive management approach and provided guidance on options appraisal, including real options analysis.

A further approach is to update engineering or design standards to account for climate change. This has been done for flood protection standards in Denmark, Germany, Australia and the United Kingdom (Wilby and Keenan, 2012). However, caution is needed to ensure that the benefits of such actions justify the additional costs, especially given future discounting of uncertain benefits and the context specific nature of adaptation.

Enabling conditions for mainstreamed decision making

Mainstreaming does not occur in a vacuum, and it is essential to understand and integrate adaptation within the existing socio-institutional landscape. For mainstreaming to be effective, entry points must be identified. Further, mainstreaming approaches need to align to the policy and institutional landscape, and consider existing processes or guidance, such as project cycle steps and appraisal documentation already in place. Since mainstreaming will be country, sector and even organisation specific, this cautions against the development of generic tools for mainstreaming adaptation.

Pragmatism is essential as any tools or guidance need to fit with the resource, time, capacity and expertise available for policy or project analysts, otherwise there is a danger that they will not get used. As an example, the UK supplementary guidance on adaptation recommends the use of real options analysis (HMT, 2009), but this proved challenging to apply outside of the context of infrastructure projects. This may mean a focus on providing information and processes that are fit for purpose, rather than perfect, particularly given the potential complexity of climate change and uncertainty analysis.

The stage at the decision-making process when adaptation is considered is critical. It is important to ensure that the mainstreaming activities come early enough in the process to influence the decision, or are targeted at key windows or "intervention opportunities" (Ballard, 2014) (see Box 6.3 for some examples). This is particularly important when there are long-lived decisions or defined policy opportunities for change.

Box 6.3. **The issue of timing and intervention opportunities**

Mainstreaming activities often need to occur early in the process if they are to be influential. An example is climate risk screening of infrastructure or investment decisions. First, there is a need for strategic issues to be identified early on, either in relation to the sector strategy or the overall investment portfolio. For example, if there are major climate risks at the river basin level, these are more difficult to address at the individual project level; instead, a strategic risk screening (e.g. within the strategic environment assessment or investment strategy) at the outset could identify such risks. Second, the analysis of climate risks and mainstreaming activities needs to occur early in the project cycle, such as at the concept or early design stage, or at the latest during the detailed design phase. The analysis should ideally be aligned to approval milestones. The inclusion of adaptation considerations at the environmental impact assessment stage, for example, is usually too late to have a major influence on project design.

For some decisions or investments, there may be very narrow and critical windows of opportunity to influence decisions. Moser and Ekstrom (2010) highlight "non-climatic windows of opportunity" such as land-use plan updates, infrastructure replacement and building renovations. Paradoxically, it may be that natural – or climate change – disasters provide opportunities to reduce further, longer-term risks since stakeholders are galvanised to act. For example, *ex post* reconstruction following a disaster such as Hurricane Katrina has resulted in an urban development patterns more cognizant of the possibility of such climate risks occurring in the future (O'Brien et al., 2012).

Source: Moser, S.C. and J. Ekstrom (2010), "A framework to diagnose barriers to climate change adaptation", Proceedings of the National Academy of Sciences of the United States of America, Vol. 107(51), *http://dx.doi.org/10.1073/pnas.1007887107*; O'Brien. K. (2012), "Global environmental change II: From adaptation to deliberate transformation", *Progress in Human Geography*, Vol. 36 (5), *http://dx.doi.org/10.1177/0309132511425767*.

In practical terms, the path from identifying potential entry points and providing tools through to implementing mainstreaming is often challenging. Achieving this requires involving a diversity of users and stakeholders, finding relevant champions, building partnerships and providing support networks. Such support networks will be particularly important as mainstreaming moves from a central unit (e.g. located in the environment ministry) out to other ministries. The complexity of decision support tools needs to be aligned to the capacity, time availability, capability and interest of sectors in switching from other tasks to climate mainstreaming. It is also critical to complement tool development with capacity development. This support must be complemented by good co-ordination and assistance. An example is the "Handbook for Local Elected Officials on Climate Change" used in Canadian municipalities to develop strong leadership on adaptation mainstreaming (Vaughan, 2012).

It is useful for decision makers to also identify opportunities that can be created by implementing adaptation, rather than focusing only on the risks and amelioration actions. For example, establishing development zones connected to ports with efficient transport may incentivise infrastructure development in areas away from the coasts (Hallegatte, 2011). The New Brunswick Climate Change Action Plan 2014-20 in Canada, for example, provides discussion of such alternatives and clearly specifies the importance of ensuring that adaptation to climate change is incorporated into every-day decisions within the province (Province of New Brunswick, 2014).

Enhancing the understanding of the barriers or constraints to adaptation can help to move from an idealised model of adaptation planning to the reality of how it plays out in practice. The UK experience provides some useful lessons (Cimato and Mullan, 2010; Frontier Economics, 2013; HMG, 2013b) These lessons include the need to identify key barriers to effective adaptation (including market, policy, behavioural and governance failures), to build organisational adaptive capacity, and to introduce enabling actions that are likely to lead to more effective adaptation (see Box 6.4 for details on each of these barriers).

Box 6.4. **Barriers to adaptation**

Barriers or constraints to adaptation are factors that make it harder to plan and implement adaptation actions (IPCC, 2014). Barriers will make adaptation less efficient or less effective. Alternatively, they may require changes that lead to missed opportunities or higher costs (Moser and Ekstrom, 2010). In 2013, the UK government in its National Adaptation Programme identified the main barriers to socially efficient adaptation as market failures, policy failures, governance failures and behavioural barriers (HMG, 2013b):

- **Market failures** can occur e.g. due to lack of information, the presence of externalities and public goods, information asymmetry and misaligned incentives. Economic theory applied to adaptation (e.g. Fankhauser, Smith and Tol, 1999; Mendhelson, 2000; Cimato and Mullan, 2010), as well as empirical observations (Mendelsohn, 2000; Osberghaus et al., 2010a; Wing and Fisher-Vanden, 2013) indicate that such actions will not receive appropriate levels of private investment. For example, Lee and Thornsbury (2010) point out that under different market structures (monopoly, oligopoly or perfect competition), the ability of investors to reap the benefits of adaptation will vary, and therefore also their incentives to invest in it.
- **Policy failures** occur when conflicting policy objectives co-exist (which is often) and there are not appropriate mechanisms for addressing these trade-offs (Frontier Economics, 2013). For example, urban development objectives may not take into account the vulnerability of assets and human systems to climatic stresses. Also, when policies result in market distortions (e.g. price or income subsidies), people will under- or over-adapt depending on how their adaptation choices will translate into income changes (Fankhauser, Smith and Tol, 1999).
- **Governance failures** refer to ineffective institutional decision-making processes, e.g. when the current structure of institutions and regulatory policies is poorly aligned to account for adaptation objectives (Craig, 2010; Spies, 2010; Stillwell et al., 2010; Stuart-Hill

Box 6.4. **Barriers to adaptation** *(cont.)*

and Schulze, 2011; Eisenack and Stecker; 2012; Huntjens et al., 2012; Herrfahrdt-Pähle, 2013). Adaptation typically requires multiple actors and institutions with different objectives, jurisdictional authority and levels of power and resources. The complexities of governance networks can constrain adaptation (see Klein et al., 2014). Overlapping mandates of government entities tend to create conflicts and slow adaptive responses. Further, lengthy bureaucratic processes and lack of transparency are an impediment to fiscal planning and access to finance, particularly relevant for developing countries (Setz et al., 2008). Poor – or lack of – leadership (Moser and Ekstrom, 2010), lack of a clear mandate, and the short-term political cycle can also represent barriers to effective decision making (Lehmann et al., 2012). Corruption within institutions also undermines adaptation efforts (Lesnikowski et al., 2013; Schilling et al., 2012).

- **Behavioural barriers** are concerned with the observed inability of individuals to take what appear to be rational decisions (i.e. to maximize their net benefits or utilities) and with their cognitive limitation in attempting to achieve their goals. This limitation manifests itself as inertia, procrastination, and the use of time-inconsistent discounting (see Simon, 1999; Jones, 1999; Cimato and Mullan, 2010). Social values and beliefs can also support or hamper adaptation (Jones and Boyd, 2011; Stafford-Smith et al., 2011; Adger et al., 2012) in so far as they frame how societies develop rules and institutions to govern risk, and to manage social change and the allocation of scarce resources (Ostrom, 2005).

Further, individuals, institutions and the natural environment will clearly adapt within the boundaries of their adaptive capacity (see Oberlack and Neumarker, 2011; Stern, 2006; Kuch and Gigli, 2007; Osberghaus et al., 2010b; Hallegatte, Lecocq and de Perthuis, 2011), and physical and biological constraints. Gender, age, education, access to infrastructure and finance, and access to markets and technology are all elements that determine the adaptive capacity of social systems. Natural systems' ability to adapt will be possible within certain climatic thresholds, and can be hampered by other non-climatic stresses (Klein et al, 2014; Cimato and Mullan, 2010), and the presence of physical barriers (e.g. the lack of corridors for species migration).

Source: Bones and Boyd (2011); Cimato and Mullan (2010); Craig (2010); Eisenack and Stecker (2012); Fankhauser, Smith and Tol (1999); Frontier Economics (2013); Hallegatte, et al. (2011); Herrfahrdt-Pähle (2013); HMG (2013b); Huntjens et al. (2012); Jones (1999); Klein et al. (2014); Kuch and Gigli (2007); Lee and Thornsbury (2010); Lehmann et al. (2012); Lesnikowski et al. (2013); Mendhelson (2000); Moser and Okstrom (2010); Oberlack and Neumarker (2011); Osberghaus et al. (2010a); Osberghaus et al. (2010b); Ostrom (2005); Schilling et al. (2012); Setz et al. (2008); Simon (1999); Spies (2010); Stafford-Smith et al. (2011); Stern (2006); Stillwell et al. (2010); Stuart-Hill and Schulze (2011); Wing and Fisher-Vanden (2013).

Recent plans for mainstreaming adaptation have started to use iterative climate risk management approaches, with examples in both OECD and non-OECD economies. Iterative approaches identify different activities compared to older adaptation impact assessment studies, with a greater focus on capacity building and enabling steps. They also provide a stronger economic justification for early intervention and longer-term action (post 2030), noting the latter is often missed in more traditional sector mainstreaming. As an example, the use of an iterative approach identified the long-term risks of climate change for the Ethiopian coffee industry, even though this was not a short-term focus for vulnerability (FDRE, 2014). Iterative approaches also provide a framework for review and evaluation. However, these approaches require detailed analysis and capacity to implement.

Decision support tools for mainstreaming

Standard public policy and project appraisal involves a systematic decision-making process: understanding the problem; identification of options; appraisal of options (and implementation approach); planning and implementation; and finally monitoring and evaluation. This approach is often formalised through guidance on impact assessment or economic appraisal and evaluation, for both policy and project decisions. Earlier impact-assessment driven studies of adaptation did not consider this broader appraisal framework, and instead largely considered adaptation, external to established processes. However, in recent years, there has been a shift towards adaptation assessment more closely aligned to the policy implementation cycle above, e.g. the UN PROVIA initiative.[2]

In terms of the standard policy or project cycle, there are two points where decision support tools are particularly important; i) for shortlisting options and ii) for prioritising the shortlisted options. The process of identifying a shortlist of options (e.g. scoping or feasibility) includes, for instance, identifying focus areas for a national plan or strategy, or a broad list of options for an individual policy or project. The prioritisation of options can be as part of detailed policy or project appraisal, based on the characteristics of climate change and other risks, as well as economic and social considerations. In the adaptation context, there is an emerging community of practice and useful examples for shortlisting and prioritisation, outlined below.

Identifying and shortlisting options

There are standard methods for shortlisting options, which may take the form of scoping economic analysis, simple attribute analysis and ranking, or stakeholder consultation and expert elicitation. These are common to most policy or project cycles, and can thus be included within a mainstreaming perspective. In the national context, most adaptation examples have been part of national or sector action planning. Similar approaches can also be taken at the project level. The aim of these processes has been to filter options down to a manageable shortlist of priorities, which can then be appraised.

Previous studies have considered a large number of methods for these early tasks, often looking at economic and other criteria to shortlist options. A key innovation of these recent approaches is that they include criteria that are of direct relevance to adaptation: for example, urgency, no- or low-regret characteristics, co-benefits, alongside the standard consideration of costs and benefits. Practical examples include the national Routeplanner exercise in the Netherlands (Van Ierland et al., 2006; De Bruin et al., 2009).

More recent assessments of adaptation actions have started to use iterative climate risk management in the scoping and initial prioritisation phase. A key advantage of this is that it identifies options with different time scales and levels of uncertainty, which can help with the phasing of responses. These include options that are beneficial, even without considering future climate change. An example is enhanced disaster risk management, which address the current adaptation deficit and helps to build future resilience. It also includes early options that build "resilience" into infrastructure development or planning processes, where there are long life-times, as well as options that introduce iterative planning and monitoring for the long-term.

Tools for appraising options

The methods outlined in the preceding paragraphs are, or can be, used to effectively incorporate climate risks into wider development policies, programmes and projects. The subsequent selection and prioritisation of shortlisted options can then be assisted by the use of a variety of decision support methods and tools. Indeed, a principal focus of policy and project analysis for adaptation to date has been on the appraisal stage, specifically the prioritisation of adaptation options. There are, however, a number of distinctive factors that are important to consider when supporting decision making on adaptation. For example, there are no simple common metrics to compare and prioritise different adaptation interventions because of the highly site- and context-specific nature of adaptation. This contrasts with mitigation options, which target a common burden (greenhouse gas emissions), that can be prioritised in terms of the cost of abating a tonne of CO_2 equivalent, using cost-effectiveness analysis. The analysis of adaptation options therefore involves additional steps to assess the impacts and to assess potential benefits, noting that many impacts are in non-market sector benefits (e.g. health, ecosystems), and that many adaptation options are non-technical in nature. There are several techniques for incorporating non-market benefits in cost-benefit analysis, but this can be a resource intensive process. This further complicates the quantification and valuation of different options.

Adaptation appraisal also has to consider the dynamic and changing nature of climate change over time, including the inter-dependencies in climate risks. The issue of future climatic (and socio-economic) uncertainty also has to be incorporated since it affects both the selection of adaptation options and the decision-framework used for prioritisation (see Chapter 4 for further discussions). Due to these diverse challenges, the most common techniques used in policy appraisal (e.g. cost-benefit analysis and cost-effectiveness analysis) have limitations in coping with adaptation. This is consistent with the latest report by the Intergovernmental Panel on Climate Change (IPCC), which highlights that "economic thinking on adaptation has evolved from a focus on cost-benefit analysis and identification of 'best economic' adaptations to the development of multi-metric evaluations" and that "economic analysis is moving away from a unique emphasis on efficiency, market solutions, and benefit-cost analysis of adaptation to include consideration of non-monetary and non-market measures; risks; inequities; behavioural biases; consideration of ancillary benefits and costs" (Chambwerra et al., 2014).

There is a growing evidence base and body of practical experience of the use of decision support approaches for adaptation appraisal. These include conventional decision-support methods, notably cost-benefit analysis, cost-effectiveness analysis and multi-criteria analysis. These also include approaches that allow for consideration of uncertainty, notably real options analysis, robust decision making, portfolio analysis and iterative risk management (see MEDIATION, 2013 and Watkiss et al., 2014 for a detailed description and review of these methods and their application to adaptation). The different methods summarised in Figure 6.3 are categorised as: traditional economic tools, uncertainty framing, and economic decision making under uncertainty. The latter two categories build on the principles in the first category and are distinct either because they introduce a dynamic component (e.g. for iterative risk management and real options analysis), they use a different/additional criterion (robust decision making or portfolio analysis), or do not rely on probabilistic data (rule-based methods).

Whilst these tools have primarily been developed in the context of project-level appraisal, in principle they can be used to inform the development of policy initiatives at the national and sectoral levels. However, these tools serve principally as an organising framework at those levels, often with semi-quantitative versions due to limited data

Figure 6.3. **Decision support tools for adaptation**

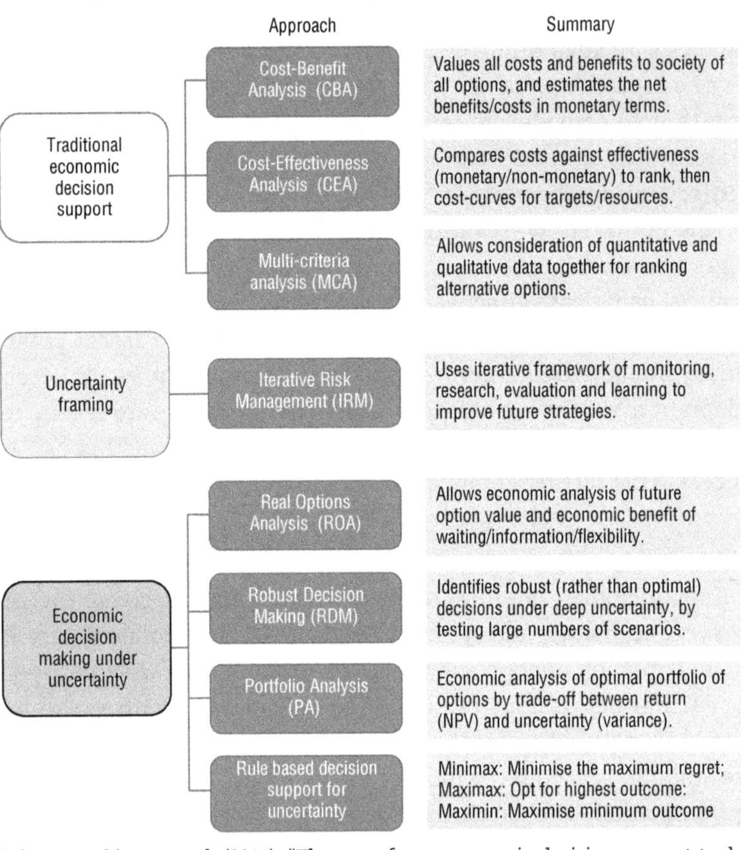

Source: Updated from Watkiss, P. et al. (2014), "The use of new economic decision support tools for adaptation assessment: A review of methods and applications, towards guidance on applicability", *Climatic Change, http://dx.doi.org/10.1007/s10584-014-1250-9.*

availability. They do, nonetheless, provide a good guide to the economic merits of the initiatives. At the project level, where data are available, they can be applied more quantitatively. Examples of adaptation applications are summarised in Table 6.1.

Table 6.1. **Examples of appraisal methods used in the adaptation context**

Tool	Published example applications
Cost-Benefit Analysis (CBA)	A South African case study examined the benefits and costs of avoiding climate change damages through structural and institutional options for increasing water supply in the Berg River Basin in the Western Cape Province (AIACC, 2006). In Germany, cost-benefit analysis was applied to assess 28 adaptation options (Tröltzsch et al., 2012).
Cost-Effectiveness Analysis (CEA)	Boyd, Wade and Walton (2006) undertook a detailed application of cost-effectiveness for water resource zones and the potential adaptation response to reduce household water deficits from future climate change in the United Kingdom. Tainio et al. (2013) investigated the cost-effectiveness of alternative conservation measures that could maintain the biodiversity of Finnish semi-natural grasslands under a changing climate.
Multi-criteria analysis (MCA)	Van Ierland et al. (2006) applied MCA to assess adaptation options for the Netherlands as part of the Routeplanner national study. This used a qualitative MCA, which included various adaptation criteria. A quantitative MCA was used in the Thames Estuary 2100 project (EA, 2009; EA, 2011) as part of a broader study looking at future coastal flood defences for London. The MCA was used to include qualitative criteria (e.g. environment and heritage) alongside formal economic cost-benefit analysis.
Real Options Analysis (ROA)	Jeuland and Whittington (2013) applied real option analysis for a water resource planning case study (large water storage projects) in Ethiopia along the Blue Nile. Van der Pol et al. (2013) looked at optimal dike investments under uncertainty with learning about increasing water levels.

Table 6.1. **Examples of appraisal methods used in the adaptation context** *(cont.)*

Tool	Published example applications
Robust Decision Making (RDM)	A comprehensive, formal application of RDM was undertaken by Lempert and Groves (2010) for Southern California's Riverside County Inland Empire Utilities Agency (IEUA). Dessai and Hulme (2007) present an example of the application for RDM to look at climate uncertainty for water supply management in the United Kingdom.
Portfolio Analysis (PO)	Crowe and Parker (2008) use portfolio analysis to investigate genetic material that could be used for the restoration or regeneration of forests under climate change futures. Hunt (2009) applied portfolio analysis to a case of local flood management in the United Kingdom, looking at portfolios of hard and soft options.
Iterative Risk Assessment (IRM)	The Thames Estuary 2100 project (EA, 2009; EA, 2011; Reeder and Ranger, 2011) developed a tidal flood-risk management adaptation plan for London using an iterative planning approach and adaptation pathways, with a detailed monitoring and evaluation strategy.

Source: ECONADAPT (2015), "The Costs and Benefits of Adaptation", results from the ECONADAPT Project, ECONADAPT consortium, *http://econadapt.eu/*.

Appraisal review findings and lessons

Analysis of conventional as well as of new appraisal support tools and discussion with practitioners reveals some interesting lessons. A recent review (ECONADAPT, 2015) sought to identify applications of new decision-support tools. It found that the number of economic applications of the new tools to adaptation remains low, in both absolute terms and relative to the use of conventional tools such as cost-benefit analysis and cost-effectiveness analysis. Applications were concentrated in sectors such as water management, coastal management and agriculture (Figure 6.4). These applications are predominantly stand-alone assessments, rather than as part of mainstreamed assessments.

Figure 6.4. **Applications of new decision support tools for adaptation**

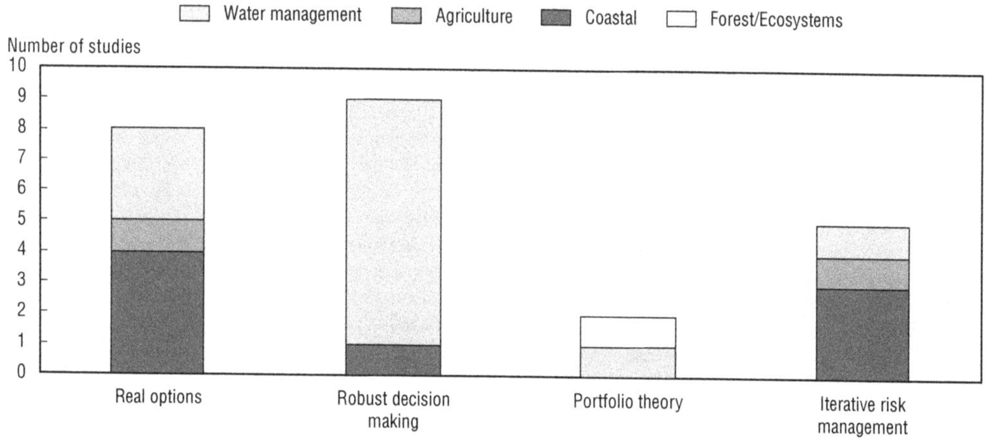

Source: ECONADAPT (2015), "The Costs and Benefits of Adaptation", results from the ECONADAPT Project, ECONADAPT consortium, *http://econadapt.eu/*.

Most economic applications are academic studies, often focused on technical adaptation, with less application in direct project or policy appraisals (ECONADAPT, 2015). The more applied studies include the application of iterative risk assessment in national policy appraisal in the Netherlands (iterative management for the Delta Programme, 2014) and in Ethiopia (the National Climate Resilience Strategy: FDRE, 2014). Iterative risk assessment has also been implemented at the project level with the application to the London Thames Estuary 2100 project (EA, 2009; EA 2011). Robust decision making has been applied to water management in the Colorado river (Groves et al., 2013), flood risk management in Ho Chi Minh

City in Viet Nam (Lempert et al., 2013) and coastal resilience planning in Louisiana (Groves and Sharon, 2013). Real options analysis has been applied in practice in the context of mitigation, but has yet to inform practical adaptation decisions. Real options analysis has also focused on sea level rise, which is easier to assess due to its slow-onset nature and known direction of change. Its advantage over cost-benefit analysis (to which it should be seen as a complement) is that it explicitly accounts for learning and adjustment of the adaptation investment over time, as climate – and other – uncertainties are resolved.

Analysis of these applications show that there are no hard or fast rules on when to use which tool, yet certain tools lend themselves more to specific contexts or sectors. The type of adaptation problem (and objective) will therefore shape the choice of the decision-support tool used. Importantly, none of these decision-support tools is universally applicable to all adaptation problems and they each have particular strengths for certain types of decisions and applications. The principal criteria on which these tools can be evaluated include: availability of probabilistic climate impact data; availability of monetary values for non-market impacts; and importance of climatic uncertainty to main decision variables. Some indicative analysis is presented in Table 6.2.

Table 6.2. **Attributes and application of decision support methods for adaptation**

Decision-support tool	Strengths	Challenges	Applicability	Potential use
Cost-Benefit Analysis (CBA)	Well known and widely applied.	Valuation of non-market sectors/ non-technical options. Uncertainty limited to probabilistic risks/ sensitivity testing.	Most useful when climate risk probabilities are known and sensitivity is small.	To identify low- and no-regret options. As a decision-support tool within iterative climate risk management
Cost-Effectiveness Analysis (CEA)	Analysis of benefits in non-monetary terms.	Single headline metric difficult to identify and less suitable for complex or cross-sectoral risks. Low consideration of uncertainty	As above, but for non-monetary sectors (e.g. ecosystems) and where social objectives exist (e.g. acceptable risks of flooding).	As above, but for market and non-market sectors where benefits are not monetised.
Multi-Criteria Analysis (MCA)	Analysis of costs and benefits in non-monetary terms.	Relies on expert judgement or stakeholders, and is subjective, including analysis of uncertainty.	When there is a mix of quantitative and qualitative data. Can include uncertainty performance as criteria.	As above, as well as for scoping options. Can complement other tools and capture qualitative aspects.
Iterative Risk Assessment (IRA)	Iterative analysis, monitoring, evaluation and learning.	Challenging when multiple risks acting together and thresholds are not always easy to identify.	Useful for long-term and uncertain challenges, especially when clear risk thresholds exist.	For appraisal over medium-longer term. Also applicable as a framework at policy level.
Real Options Analysis (ROA)	Value of flexibility, information.	Requires economic valuation (see CBA), probabilities and clear decision points.	Large irreversible decisions, where information is available on climate risk probabilities.	Economic analysis of major capital investment decisions. Analysis of flexibility within major projects.
Robust Decision Making (RDM)	Robustness rather than optimisation.	High computational analysis (formal) and large number of model runs.	When uncertainty is large. Can use a mix of quantitative and qualitative information.	Identifying low- and no-regret options and robust decisions for investments with long life-times.
Portfolio Analysis (PA)	Analysis of portfolios rather than individual options.	Requires economic data and probabilities. Issues of interdependence.	When number of complementary adaptation actions and good information.	Project based analysis of future combinations. Designing portfolio mixes as part of iterative pathways.

Source: Adapted from Watkiss, P. et al. (2014), "The use of new economic decision support tools for adaptation assessment: A review of methods and applications, towards guidance on applicability", *Climatic Change*, *http://dx.doi.org/10.1007/s10584-014-1250-9*.

As mentioned above, policy-level assessments are more likely to make use of the established tools that provide a framework for more aggregated analysis. Nonetheless, iterative risk frameworks and robust decision making also have high potential for programme and sector analysis though they are more widely used at the project level. At the project level, tool selection will be influenced by data availability and the level of uncertainty. Several approaches, such as real options and portfolio theory, require subjective probabilistic inputs at a minimum, but perform better when objective probabilistic inputs are available. Suitable probabilistic climate projections are not available for all geographic regions.

While existing decision support tools, including cost-benefit analysis, can be used for studies that are focused on current climate variability, adaptation interventions are often in areas that are difficult to value, and usually involve a lack of quantitative information. In such cases, cost-effectiveness, multi-attribute analysis (or multi-criteria analysis) may be more practical, notwithstanding the limitations of these approaches. For the analysis of short-term decisions with long life-times and longer-term challenges, a greater focus on new decision support tools is warranted. Robust decision making has broad application for current and future time periods. When investments are nearer term (especially high up-front capital investments), and where there is an existing adaptation deficit, real options analysis is a potentially useful tool. For long-term applications in conditions of low current adaptation deficit, iterative risk assessment may be more applicable. Importantly, while the tools are presented individually, they are not mutually exclusive.

It is worth noting that the differences between the tools are not limited to data and capacity constraints but may have a material impact on the order of prioritisation of adaptation options. Klijn, Mens and Asselman (2014) demonstrate that applying robust decision making results in a different order from cost-benefit analysis, and cost-benefit analysis produces a different order from cost-effectiveness analysis.

A key message, however, is that all the new methods are resource-intensive and can be technically complex. This constrains their formal application to large investment decisions or major risks, which are priority projects for adaptation. These issues are likely to limit future application in the mainstreaming context, especially in developing countries, but also in OECD countries, as shown by early implementation experience (HMT, 2008). The translation into sectoral contexts, with analysts who may not have extensive knowledge of climate projections and uncertainty, is likely to be difficult.

A critical question is therefore whether the concepts in these approaches to uncertainty can be used in "light-touch" approaches that capture their conceptual aspects, while maintaining a degree of economic rigour, both at policy and project level. This would allow a wider application in qualitative or semi-quantitative analysis. Possible elements include the broad use of decision tree structures from real options analysis, the concepts of robustness testing from robust decision making, the shift towards portfolios of options from portfolio analysis, and the focus on evaluation and learning from iterative risk assessment for long-term strategies. There has been some progress advancing these types of "light-touch" applications (Hallegatte et al., 2012; Ranger et al., 2013). However, more research needs to be undertaken to better understand how and where the trade-offs between quantitative analysis and pragmatic application can be made.

Additional issues with mainstreaming and appraisal

The discussion and the tools above are predominantly focused on the challenges of timing and uncertainty. However, as outlined in Chapters 3 and 4, there are a number of other methodological challenges with adaptation. These include assumptions regarding the choice of discount rate, equity weighting or distributional issues and risk preferences. They also include issues of baselines, analysis of non-technical options, issues of scale and aggregation, and transferability of benefits and costs. These challenges potentially apply at both the policy and programme levels, as well as the project-level. The larger scale and the increased number of actors may also mean that the appraisal and implementation processes and the monitoring of outcomes become more challenging, with differing views on how to best address these issues.

These additional issues can potentially complicate the process of mainstreaming. For instance, discounting means that longer-term benefits have less weight in the decision-making process than those occurring now. While this takes account of preferences, and allows the effective allocation of resources, it can reduce the attractiveness of more sustainable options, for example, it usually takes several years for the benefits of sustainable agriculture to become apparent (Berger and Chambwera, 2010). Discounting can make it difficult to justify early action to address longer-term major impacts. While declining discount rates are already recommended for use in some OECD countries, they are likely to have relatively modest impacts compared to the use of constant values. The shift to iterative risk management, and the consideration of the value of information, learning and option values, along with analysis of robustness and flexibility, helps to capture the economic justification for appropriate early actions.

Box 6.5. **Discounting and discount rates**

Discounting is used to compare costs and benefits arising at different point in time. The higher the discount rate, the lower the weight placed on costs and benefits arising in the future.

The market rate(s) of interest will be the most relevant in undertaking financial appraisal of projects by the private sector. The market rate of interest arises because individuals attach less weight to a benefit or cost in the future than they do to a benefit or cost now. Impatience, or "pure time preference", is one reason why the present is preferred to the future. Historically, 6-20% effectively represents a typical range for the market rates of interest – the lower end of the range being more common in OECD countries and the higher end being common in less developed countries.

The social discount rate attempts to measure the rate at which the utility of consumption falls over time for society and is therefore distinct from the market interest rate, which is determined by individual preferences expressed in financial markets. Social discount rates, as used by the public sector, tend to be lower than market interest rates – a range of 3% to 12% is typical.

It should be noted that when long time periods need to be considered, as in the context of climate change, the effect of discounting is to weight present values over future values so that the damages associated with climate change become very small. For example, with a horizon of around 100 years, a discount rate of 4 percent implies that damage of USD 100 at the end of the period is valued at USD 8 today.

Some conclusions can be drawn regarding the practical consequences for discounting as applied to adaptation. First, market rates will apply to those investment decisions that are made by private sector actors. These rates may be expected to constrain investments in adaptation to climate change.

Second, where adaptation investments in the public sector are likely to have lifetimes of less than 30-40 years, standard public sector discount rates will be appropriate. Consequently, and as with private sector adaptation, near-term climate change impacts will be most important in determining the selection of adaptation options. This outcome is likely to be exacerbated by the fact that impacts in the distant future are much more uncertain, so reducing the influence they have on adaptation option selection.

Third, investments in adaptation that have long life-times (for example, over 40 years) may be more appropriately discounted at declining rates, as in the example given above. However, for the sake of consistency, it is important that other, non-adaptation, public sector investments are subject to the same discounting profile.

A further issue relates to the distributional patterns of climate change, which disproportionately affects poor and vulnerable groups (IPCC, 2014). Some studies highlight that, in the context of adaptation, cost-benefit analyses do not always capture inequality and impacts on the most vulnerable, as it focuses on more valuable assets and groups (e.g. Cartwright et al., 2013). There is also a question of how costs and benefits can be balanced where one community or stakeholder benefits and another loses. It is possible to take these factors into account in appraisal by using equity (distributional) weights or looking at the distributional effects of policies or projects, and it may be advantageous to use these more frequently than is currently the case. Finally, it remains an open question as to what extent the costs of adaptation (e.g. a flood protection scheme) should be borne by the beneficiaries of the measure and to what extent they should be shared across a larger population group (Fankhauser and Soure, 2012). Different societies will come to different conclusions based on their social welfare function.

The challenges above are not unique to adaptation. This is because the concepts of mainstreaming would suggest compatibility with existing systems, processes and guidance. Most policy and project appraisal guidance has recommended approaches for baseline setting, and the analysis of non-market and qualitative benefits. Further, while discount rates and equity weights are potentially a contentious issue, existing guidance and decisions on these will already exist in the policy or project appraisal context. To illustrate, if existing appraisal practice recommends declining discount rates (e.g. HMT, 2009) or intergenerational rates (HMT, 2008), then these should also be applied to adaptation decisions: if not, then it is unlikely that adaptation alone will lead to a change of practice. Similarly, if equity weights are not currently used, e.g. in existing cost-benefit analysis, then it is unlikely that adaptation will be the trigger to change organisational practice and the current practice of qualitative analysis of distributional consequence of policies is likely to continue. However, as more experience of adaptation mainstreaming develops, it may be necessary to re-examine existing systems, to see if they adequately address these issues.

Notes

1. The views expressed in this publication are the sole responsibility of the authors and do not necessarily reflect the views of the European Commission. The European Community is not liable for any use made of this information. The project and this review also benefited from co-funding from Canada's International Development Research Centre, as part of the project "The Economics of Adaptation and Climate-Resilient Development" – however the views expressed are entirely those of the study team and do not necessarily reflect the views of IDRC.
2. Programme of Research on Climate Change Vulnerability, Impacts and Adaptation (PROVIA) is a global initiative which aims to provide direction and coherence at the international level for research on vulnerability, impacts and adaptation. Provia was supported by the Mediation Project (Methodology for Effective Decision-making on Impacts and AdaptaTION). This project provided scientific and technical information about climate change impacts, vulnerability and adaptation options, including the adaptation learning cycle, methods, decision support and information (Hinkel and Bisaro, 2013).

References

Adger, W.N. et al. (2012), "Cultural dimensions of climate change impacts and adaptation", *Nature Climate Change*, Vol. 3, *http://dx.doi.org/10.1038/nclimate1666.*

AfDB (2011), *Climate Screening and Adaptation Review & Evaluation Procedures*, African Development Bank Group, Tunis.

Agrawala, S. et al. (2010), "Incorporating Climate Change Impacts and Adaptation in Environmental Impact Assessments: Opportunities and Challenges", *OECD Working Paper, http://dx.doi.org/10.1787/19970900.*

AIACC (Assessments of Impacts and Adaptations to Climate Change) (2006), "Estimating and Comparing Costs and Benefits of Adaptation Projects: Case Studies in South Africa and The Gambia", Final Report Submitted to AIACC, the International START Secretariat, Washington, DC.

Ballard, D. (2014), "CAFCA – Climate Adaptation Financing Coastal Areas", *Method Paper.*

Berger and Chambwera (2010), "Beyond cost-benefit: Developing a complete toolkit for adaptation decisions", *IIED Briefing Note*, International Institution for Environment and Development, London.

Boyd, R., S. Wade and H. Walton (2006), *Climate Change Impacts and Adaptation: Cross-Regional Research Project(E)*, Defra, UK.

de Bruin, K. et al. (2009), "Adapting to climate change in the Netherlands: An inventory of climate adaptation options and ranking of alternatives", *Climatic Change*, Vol. 95(1-2), *http://dx.doi.org/10.1007/s10584-009-9576-4.*

Cartwright, A. et al. (2013), "Economics of climate change adaptation at the local scale under conditions of uncertainty and resource constraints: the case of Durban, South Africa", *Environment and Urbanization*, No. 6, *http://dx.doi.org.10.1177/0956247813477814.*

Chambwera, M. et al. (2014), Economics of adaptation. In: *Climate Change 2014: Impacts, Adaptation, and Vulnerability. Part A: Global and Sectoral Aspects.* Contribution of Working Group II to the Fifth Assessment Report of the Intergovernmental Panel on Climate Change. Cambridge University Press, Cambridge, United Kingdom and New York, NY, USA, pp. 945-977.

Craig, R.K. (2010), "'Stationarity is Dead' – Long Live Transformation: Five Principles for Climate Change Adaptation Law", *Harvard Environmental Law Review*, Vol. 34(1).

Crowe K.A. and W.H.Parker (2008), "Using portfolio theory to guide reforestation and restoration under climate change scenarios", *Climatic Change*, Vol. 89, *http://dx.doi.org/10.1007/s10584-007-9373-x.*

Cimato, F. and M. Mullan (2010), "Adapting to Climate Change: Analysing the Role of Government", *Defra Evidence and Analysis Series*, Paper No. 1, Defra, London.

de Bruin, K. et al. (2009), "Adapting to climate change in The Netherlands: an inventory of climate adaptation options and ranking of alternatives", *Climatic Change*, Vol. 95(1-2), *http://dx.doi.org/10.1007/s10584-009-9576-4.*

Deltacommissie (2008), *Working together with water. A living land builds for its future*, Findings of the Deltacommissie, DeltaCommissie, The Hague.

Delta Programme (2014), *Working on the delta: Promising solutions for tasking and ambitions*, The Delta Programme, The Hague.

Delta Programme (2011), *Working on the delta: Investing in a safe and attractive Netherlands, now and in the future*, The Delta Programme, The Hague.

Dessai S. and M. Hulme (2007), "Assessing the robustness of adaptation decisions to climate change uncertainties: A case study on water resources management in the East of England", *Global Environmental Change*,Vol. 17(1), *http://dx.doi.org/10.1016/j.gloenvcha.2006.11.005.*

DFID (2014), *Early Value-for-Money Adaptation: Delivering VfM Adaptation using Iterative Frameworks and Low-Regret Options*, UK Department for International Development, London.

Downing, T.E. (2012), "Views of the frontiers in climate change adaptation economics", *Wiley Interdisciplinary Reviews: Climate Change*, Vol. 3(2), *http://dx.doi.org/10.1002/wcc.157.*

EA (2011), "Thames Estuary 2100: Strategic Outline Programme", Environment Agency (UK), Bristol.

EA (2009), "Thames Estuary 2100 Plan Technical Report – Appendix H: Appraisal in TE2100", Environment Agency (UK), Bristol.

ECONADAPT (2015), "The Costs and Benefits of Adaptation", results from the ECONADAPT Project, ECONADAPT consortium, *http://econadapt.eu/.*

EEA (2014), *National adaptation policy processes in European countries: 2014*, European Environment Agency, Copenhagen.

Eisenack, K. and R. Stecker (2012), A framework for analyzing climate change adaptations as actions. *Mitigation and Adaptation Strategies for Global Change*, 17, 243-260.

Fankhauser, S. and R. Soare (2012), "Strategic Adaptation to Climate Change in Europe", *EIB Working Papers* 2012 / 01.

Fankhauser, S., J.B. Smith and R.S.J. Tol (1999), "Weathering climate change: Some simple rules to guide adaptation decisions", *Ecological Economics*, Vol. 30(1), *http://dx.doi.org/10.1016/S0921-8009(98)00117-7*.

Frontier Economics (2013), "The Economics of Climate Resilience CA0401", Frontier Economics Ltd, Ecofys UK Ltd, Irbaris LLP. *http://randd.defra.gov.uk/Default.aspx?Module=More&Location=None&ProjectID=18016* (accessed 16 March 2015).

Groves, D.G. and C. Sharon (2013), "Planning Tool to Support Planning the Future of Coastal Louisiana", *Journal of Coastal Research*, Special Issue 67 – Louisiana's 2012 Coastal Master Plan Technical Analysis, *http://dx.doi.org/10.2112/SI_67_10*.

Groves, D.G. et al. (2013), *Adapting to a Changing Colorado River. Making Future Water Deliveries More Reliable Through Robust Management Strategies*, Prepared for the US Bureau of Reclamation and conducted in the Environment, Energy and Economic Development Program within RAND Justice, Infrastructure and Environment, Santa Monica.

Haasnoot, M. et al. (2013), "Dynamic adaptive policy pathways: A method for crafting robust decisions for a deeply uncertain world", *Global Environmental Change*, Vol. 23(2), *http://dx.doi.org/10.1016/j.gloenvcha.2012.12.006*.

Hallegatte, S. (2009), "Strategies to adapt to an uncertain climate change", *Global Environmental Change*, Vol. 19(2), *http://dx.doi.org/10.1016/j.gloenvcha.2008.12.003*.

Hallegatte, S. (2011), "How Economic Growth and Rational Decisions Can Make Disaster Losses Grow Faster Than Wealth", *Policy Research Working Paper* 5617, World Bank, Washington, DC.

Hallegatte, S. et al. (2012), "Investment Decision Making Under Deep Uncertainty: Application to Climate Change", *Policy Research Working Papers*, No. 6193, World Bank, Washington, DC, *http://dx.doi.org/10.1596/1813-9450-6193*.

Hallegatte, S., F. Lecocq and C. de Perthuis (2011), "Designing climate change adaptation policies: An economic framework", *Policy Research Working Paper Series*, No. 5568, World Bank, Washington, DC, *http://dx.doi.org/10.1596/1813-9450-5568*.

Herrfahrdt-Pähle, E. (2013), "Integrated and adaptive governance of water resources: the case of South Africa", *Regional Environmental Change*, Vol. 13(3), *http://dx.doi.org/10.1007/s10113-012-0322-5*.

Hinkel, J. and A. Bisaro (2013), "Methodological choices in problem-oriented adaptation research: a diagnostic framework", *Regional Environmental Change*, MEDIATION Special Issue.

HMG (2013a), *The National Adaptation Programme: Making the country resilient to a changing climate*, HM Government, London.

HMG (2013b), *The National Adaptation Programme Report: Analytical Annex*, Economics of the National Adaptation Programme, HM Government, London.

HMT (2009), *Accounting for the Effects of Climate Change: Supplementary Green Book guidance*, HM Treasury, London.

HMT (2008), *Intergenerational wealth transfers and social discounting: Supplementary Green Book guidance*, HM Treasury, London.

Hunt, A. (2009), "Economic Aspects of Climate Change Impacts and Adaptation in the UK", PhD Thesis, University of Bath.

Hunt, A. and P. Watkiss (2011), "Climate change impacts and adaptation in cities: a review of the literature", *Climatic Change*, Vol. 104(1), *http://dx.doi.org/10.1007/s10584-010-9975-6*.

Huntjens, P. et al. (2012), "Institutional design propositions for the governance of adaptation to climate change in the water sector", *Global Environmental Change*, Vol. 22(1), *http://dx.doi.org/10.1016/j.gloenvcha.2011.09.015*.

IPCC (Intergovernmental Panel on Climate Change) (2014), *Climate Change 2014: Impacts, Adaptation, and Vulnerability*, Contribution of Working Group II to the Fifth Assessment Report of the Intergovernmental Panel on Climate Change, Cambridge University Press, Cambridge and New York.

Jeuland, M. and D. Whittington (2013), "Water Resources Planning under Climate Change: A 'Real Options' Application to Investment Planning in the Blue Nile", *Environment for Development. Discussion Paper Series*, No. 13-5.

Jones, B.D. (1999), "Bounded Rationality", *Annual Review of Political Science*, Vol. 2: 297-321, *http://dx.doi.org/10.1146/annurev.polisci.2.1.297*.

Jones, L. and E. Boyd (2011), "Exploring social barriers to adaptation: insights from Western Nepal", *Global Environmental Change*, Vol. 21(4), *http://dx.doi.org/10.1016/j.gloenvcha.2011.06.002*.

Frontier Economics (2013), "The Economics of Climate Resilience CA0401: Synthesis Report", Frontier Economics, IRBARIS and Ecofys, London.

Klijn, F., M.J.P. Mens and N.E.M. Asselman (2014), "Robustness analysis for flood risk management planning: On risk-based decision making beyond simple economic reasoning, exemplified for the Meuse River (Netherlands), *Paper presented at the 6th International Conference of Flood Management*, September 2014, Sao Paulo.

Klein, R.J.T. et al., (2014), Adaptation opportunities, constraints, and limits. In: *Climate Change 2014: Impacts, Adaptation, and Vulnerability. Part A: Global and Sectoral Aspects*. Contribution of Working Group II to the Fifth Assessment Report of the Intergovernmental Panel on Climate Change.

Klein, R.J.T. and Å. Persson (2008), "Financing Adaptation to Climate Change: Issues and Priorities", European Climate Platform (ECP), *Report* No. 8.

Kuch, P.J. and S. Gigli (2007), *Economic Approaches to Climate Change Adaptation: And their role in project prioritisation and appraisal*, GTZ (ed.), Climate Protection Programme in Developing Countries, Eschborn.

Least Developed Countries (LDC) Expert Group, 2012, *National Adaptation Plans, Technical guidelines for the national adaptation plan process*, Bonn: UNFCCC secretariat, Bonn, Germany, December 2012. Available at: *http://unfccc.int/NAP*.

Lee S. and S. Thornsbury (2010), "The effect of market structure on adaptation to climate change in agriculture", Contributed paper at the IATRC Public Trade Policy Research and Analysis Symposium "Climate Change in World Agriculture: Mitigation, Adaptation, Trade and Food Security", June 27-29, Universität Hohenheim, Stuttgart.

Lehmann, P. et al. (2012), "Understanding barriers and opportunities for adaptation planning in cities", *UFZ Discussion Papers*, No. 19/2012.

Lempert, R.J. et al. (2013), "Ensuring Robust Flood Risk Management in Ho Chi Minh City", *Policy Research Working Papers*, No. 6465, World Bank, Washington, DC, *http://dx.doi.org/10.1596/1813-9450-6465*.

Lempert, R.J. and D.G. Groves (2010), "Identifying and evaluating robust adaptive policy responses to climate change for water management agencies in the American west", *Technological Forecasting and Social Change*, Vol. 77(6), *http://dx.doi.org/10.1016/j.techfore.2010.04.007*.

Lesnikowski, A.C. et al. (2013), "National-level factors affecting planned, public adaptation to health impacts of climate change", *Global Environmental Change*, 23(5), *http://dx.doi.org/10.1016/j.gloenvcha.2013.04.008*.

Li, J., M. Mullan and J.Helgeson (2014), "Improving the practice of economic analysis of climate change adaptation", *Journal of Benefit Cost Analysis*, Vol. 5(3), *http://dx.doi.org/10.1515/jbca-2014-9004*.

McGray, H., A. Hammill and R. Bradley (2007), *Weathering the Storm: Options for Framing Adaptation and Development*, World Resources Institute, Washington, DC.

MEDIATION (2013), Decision Support Methods for Climate Change Adaption, *Technical Policy Briefing Notes*, MEDIATION project (Methodology for Effective Decision-making on Impacts and AdaptaTION). Available at *http://mediation-project.eu/output/technical-policy-briefing-notes* (Accessed, March 2015).

Mendelsohn, R. (2000), "Efficient adaptation to climate change", *Climatic Change*, 45,583-600.

Moser, S.C. and J. Ekstrom (2010), "A framework to diagnose barriers to climate change adaptation", *Proceedings of the National Academy of Sciences of the United States of America*, Vol. 107(51), *http://dx.doi.org/10.1073/pnas.1007887107*.

Mullan, M., et al. (2013), "National Adaptation Planning: Lessons from OECD Countries", *OECD Environment Working Papers*, No. 54, OECD Publishing, Paris, *http://dx.doi.org/10.1787/5k483jpfpsq1-en*.

Nunan, F., A. Campbell and E. Foster (2012), *Environmental Mainstreaming: The Organisational Challenges of Policy Integration*, Public Administration and Development 32, 262-277 *http://dx.doi.org/10.1002/pad.1624*.

Oberlack, C. and B. Neumarker, (2011), "Economics, institutions and adaptation to climate change", *The Constitutional Economics Network Working Papers*, No. 04-2011.

O'Brien. K. (2012), "Global environmental change II: From adaptation to deliberate transformation", *Progress in Human Geography*, Vol. 36 (5), *http://dx.doi.org/10.1177/0309132511425767*.

OECD (Organisation for Economic Co-ordination and Development) (2012), *Greening Development: Enhancing Capacity for Environmental Management and Governance*, OECD Publishing, Paris, *http://dx.doi.org/10.1787/9789264167896-en*.

OECD (2009), *Integrating Climate Change Adaptation into Development Co-operation: Policy Guidance*, OECD Publishing, Paris, *http://dx.doi.org/10.1787/9789264054950-en*.

Osberghaus, D. et al. (2010a), "The role of the government in adaptation to climate change", *Environment and Planning C: Government and Policy*, Vol. 28(5), *http://dx.doi.org/10.1068/c09179j*.

Osberghaus et al. (2010b), "Individual Adaptation to Climate Change: The Role of Information and Perceived Risk", *ZEW Discussion Paper No. 10-061*, Zentrum für Europäische Wirtschaftsforschung (ZEW) GmbH, Centre for European Economic Research, Mannheim, Germany.

Stern, N. et al. (2006), *The Economics of Climate Change*, HM Treasury, London.

Ostrom, E. (2005), *Understanding institutional diversity*, Princeton University Press, Princeton.

Province of New Brunswick (2014), *New Brunswick Climate Change Action Plan 2014-2020*, Province of New Brunswick.

Ranger, N., T. Reeder and J. Lowe (2013), "Addressing 'deep' uncertainty over long-term climate in major infrastructure projects: four innovations of the Thames Estuary 2100 Project", *European Journal of Decision Process*, 1:233-262, *http://dx.doi.org/10.1007/s40070-013-0014-5*.

Ranger, N. et al. (2010), "Adaptation in the UK: a decision-making process", *Policy Brief*, Grantham Research Institute on Climate Change and the Environment and Centre for Climate Change Economics and Policy.

Reeder, T. and N. Ranger (2011), "How do you adapt in an uncertain world? Lessons from the Thames Estuary 2100 project", *World Resources Report Uncertainty Series*, WRI, Washington, DC.

Republic of Rwanda (2012), *Economic Development And Poverty Reduction Strategy 2013-2018*, Published by the Republic of Rwanda, Kigali, Rwanda.

Schilling, J. et al. (2012), "Climate change, vulnerability and adaptation in North Africa with focus on Morocco", *Agriculture, Ecosystems and Environment*, Vol. 156, *http://dx.doi.org/10.1016/j.agee.2012.04.021*.

Simon, H.A. (1999), "The potlatch between economics and political science", in Alt, J.E., M. Levi and E. Ostrom (eds.), *Competition and cooperation: Conversations with Nobelists about economics and political science*, Cambridge University Press, Cambridge.

Spies, T.A. et al. (2010), "Climate change adaptation strategies for federal forests of the Pacific Northwest, USA: Ecological, policy, and socio-economic perspectives", *Landscape Ecology*, Vol. 25(8), *http://dx.doi.org/10.1007/s10980-010-9483-0*.

Stafford-Smith, M. et al. (2011), "Rethinking adaptation for a 4 °C world", *Philosophical Transactions of the Royal Society*, Vol. 369, *http://dx.doi.org/10.1098/rsta.2010.0277*.

Stern N. et al. (2006), *The Economics of Climate Change*, HM Treasury, London.

Stillwell, A.S., M.E. Clayton and M.E. Webber (2011), "Technical analysis of a river basin-based model of advanced power plant cooling technologies for mitigating water management challenges", *Environmental Research Letters*, Vol. 6, *http://dx.doi.org/10.1088/1748-9326/6/3/034015*.

Stuart-Hill, S. and R.E. Schulze (2010), "Does South Africa's water law and policy allow for climate change adaptation?", *Climate and Development*, Vol. 2(2), *http://dx.doi.org/10.3763/cdev.2010.0035*.

Tainio, A. et al. (2014), "Conservation of grassland butterflies in Finland under a changing climate", *Regional Environmental Change*, *http://dx.doi.org/10.1007/s10113-014-0684-y*.

Tröltzsch, J., H. Lückge and C. Sartorius (2012), *Kosten und Nutzen von Anpassungsmaßnahmen an den Klimawandel: Analyse von 28 Anpassungsmaßnahme in Deutschland* [Costs and benefits of climate adaptation measures. Analysis of 28 adaptation measures in Germany], *Climate Change*, No. 10/2012, UBA, Dessau.

True, J. (2010), *Mainstreaming Gender in International Relations. Gender Matters in Global Politics* (New York: Routledge), L.J. Shepherd (ed.), pp. 189-203.

UNDP (United Nations Development Programme)/UNEP (United Nations Environment Programme) (2011), *Mainstreaming Climate Change Adaptation into Development Planning: A Guide for Practitioners*, UNDP-UNEP Poverty-Environment Initiative, Nairobi.

UNFCCC (2009), "Potential costs and benefits of adaptation options: A review of existing literature", *FCCC/TP/2009/2*, United Nations Framework Convention on Climate Change, Bonn.

van der Pol, T.D., E.C. van Ierland and H.P. Weikard (2013), "Optimal dike investments under uncertainty and learning about increasing water levels", *Journal of Flood Risk Management*, Vol. 7(4), *http://dx.doi.org/10.1111/jfr3.12063*.

van Ierland, E.C. et al. (2006), *A qualitative assessment of climate change adaptation options and some estimates of adaptation costs*, Routeplanner subprojects 3, 4 and 5, Wageningen UR.

Vaughan, H. (2012), *Leadership & Legacy: Handbook for Local Elected Officials on Climate Change*, ICLEI-Local Governments for Sustainability (Management) Inc.

Watkiss, P. (2014), *Design of Policy-Led Analytical Framework for the Economics of Adaptation*, Deliverable 1.2 of the ECONADAPT project. ECONADAPT, *http://econadapt.eu/*.

Watkiss, P. and A. Hunt (2011), "Method for the UK Adaptation Economic Assessment" (Economics of Climate Resilience), *Final Report to Defra*, May, Deliverable 2.2.1.

Watkiss, P. et al. (2014), "The use of new economic decision support tools for adaptation assessment: A review of methods and applications, towards guidance on applicability", *Climatic Change*, *http://dx.doi.org/10.1007/s10584-014-1250-9*.

Wilby, R.L. and S. Dessai (2010), "Robust adaptation to climate change", *Weather*, Vol. 65(7), *http://dx.doi.org/10.1002/wea.543*.

Wilby, R.L and R. Keenan (2012), "Adapting to flood risk under climate change", *Progress in Physical Geography*, Vol. 36(3), *http://dx.doi.org/10.1177/0309133312438908*.

Wilby, R.L. (2012), "Frameworks for delivering regular assessments of the risks and opportunities from climate change: An independent review of the first UK Climate Change Risk Assessment", *Final Report to the Committee on Climate Change*, Loughborough University.

Wing, I.S. and K. Fisher-Vanden (2013), "Confronting the challenge of integrated assessment of climate adaptation: a conceptual framework", *Climatic Change*, 117(3), *http://dx.doi.org/10.1007/s10584-012-0651-x*.

ORGANISATION FOR ECONOMIC CO-OPERATION AND DEVELOPMENT

The OECD is a unique forum where governments work together to address the economic, social and environmental challenges of globalisation. The OECD is also at the forefront of efforts to understand and to help governments respond to new developments and concerns, such as corporate governance, the information economy and the challenges of an ageing population. The Organisation provides a setting where governments can compare policy experiences, seek answers to common problems, identify good practice and work to co-ordinate domestic and international policies.

The OECD member countries are: Australia, Austria, Belgium, Canada, Chile, the Czech Republic, Denmark, Estonia, Finland, France, Germany, Greece, Hungary, Iceland, Ireland, Israel, Italy, Japan, Korea, Luxembourg, Mexico, the Netherlands, New Zealand, Norway, Poland, Portugal, the Slovak Republic, Slovenia, Spain, Sweden, Switzerland, Turkey, the United Kingdom and the United States. The European Commission takes part in the work of the OECD.

OECD Publishing disseminates widely the results of the Organisation's statistics gathering and research on economic, social and environmental issues, as well as the conventions, guidelines and standards agreed by its members.

ÉDITIONS OCDE, 2, rue André-Pascal, 75775 PARIS CEDEX 16
(97 2015 08 1 P) ISBN 978-92-64-23460-4 – 2015